Invention and Innovation

Invention and Innovation

The Social Context of Technological Change 2:
Egypt, the Aegean and the Near East,
1650–1150 BC

*Proceedings of a conference held at the
McDonald Institute for Archaeological Research,
Cambridge, 4–6 September 2002*

Edited by
Janine Bourriau and Jacke Phillips

Oxbow Books

First published in the United Kingdom in 2004. Reprinted in 2016 by
OXBOW BOOKS
10 Hythe Bridge Street, Oxford OX1 2EW

and in the United States by
OXBOW BOOKS
1950 Lawrence Road, Havertown, PA 19083

© Oxbow Books and the individual authors, 2016

Paperback Edition: ISBN 978-1-84217-150-9
Digital Edition: ISBN 978-1-78570-420-8 (ePub)

A CIP record for this book is available from the British Library

All rights reserved. No part of this book may be reproduced or transmitted in any form or by any means, electronic or mechanical including photocopying, recording or by any information storage and retrieval system, without permission from the publisher in writing.

For a complete list of Oxbow titles, please contact:

UNITED KINGDOM	UNITED STATES OF AMERICA
Oxbow Books	Oxbow Books
Telephone (01865) 241249, Fax (01865) 794449	Telephone (800) 791-9354, Fax (610) 853-9146
Email: oxbow@oxbowbooks.com	Email: queries@casemateacademic.com
www.oxbowbooks.com	www.casemateacademic.com/oxbow

Oxbow Books is part of the Casemate Group

Printed and bound in Great Britain by
Marston Book Services Ltd, Oxfordshire

Front cover: Faience Argonaut from the Palace of Zakros, Crete

Contents

List of contributors
Preface and Acknowledgements

1. Hopeful Monsters? Invention and Innovation in the Archaeological Record *(Andrew Shortland)* 1

2. Identity and Occupation: how did individuals define themselves and their work in the Egyptian New Kingdom? *(Ian Shaw)* 12

3. Canaan in Egypt: archaeological evidence for a social phenomenon *(Rachael Thyrza Sparks)* 25

4. The Provenance of Canaanite Amphorae found at Memphis and Amarna in the New Kingdom: results 2000–2002 *(Laurence Smith, Janine Bourriau, Yuval Goren, Michael Hughes and Margaret Serpico)* 55

5. The Beginnings of Amphora Production in Egypt *(Janine Bourriau)* 78

6. Natural Product Technology in New Kingdom Egypt *(Margaret Serpico)* 96

7. Minoan and Mycenaean Technology as Revealed Through Organic Residue Analysis *(Holley Martlew)* 121

8. The Production Technology of Aegean Bronze Age Vitreous Materials *(Marina Panagiotaki, Yannis Maniatis, Despina Kavoussanaki, Gareth Hatton and Mike Tite)* 149

9. Egyptian Sculptors' Models: functions and fashions in the 18th Dynasty *(Sally-Ann Ashton)* 176

10. How to Build a Body Without One: composite statues from Amarna *(Jacke Phillips)* 200

List of Contributors

SALLY-ANN ASHTON
Fitzwilliam Museum
Trumpington Street
Cambridge CB2 1RB
UK
sa337@cam.ac.uk

JANINE BOURRIAU
McDonald Institute for Archaeological
 Research
Downing Street
Cambridge CB2 3ER
UK
jdb29@cam.ac.uk

YUVAL GOREN
Department of Archaeology and
 Ancient Near Eastern Cultures
Tel-Aviv University
Tel-Aviv 69978
Israel
ygoren@post.tau.ac.il

GARETH HATTON
Research Laboratory for Archaeology
 and the History of Art
6 Keble Road
Oxford OX1 3QJ
UK
gareth.hatton@rlaha.ox.ac.uk

MICHAEL HUGHES
4 Welbeck Rise
Harpenden
Herts. AL5 1SL
UK
michaelhughes@archsci.freeserve.co.uk

DESPINA KAVOUSSANAKI
Laboratory for Archaeometry
Institute of Materials
NCSR "Demokritos"
15310 Ag. Paraskevi
Attikis, Greece

YANNIS MANIATIS
Laboratory for Archeometry
Institute of Materials
NCSR "Demokritos"
15310 Ag. Paraskevi
Attikis, Greece
maniatis@ims.demokritos.gr

HOLLEY MARTLEW
The Holley Martlew Archaeological
 Foundation
Tivoli House
Tivoli Road
Cheltenham
Glos. GL50 2TD
UK

MARINA PANAGIOTAKI
Department of Mediterranean Studies
University of the Aegean
1, Demokratias Street
85100 Rhodes
Greece
mpanagiotaki@her.forthnet.gr

JACKE PHILLIPS
McDonald Institute for Archaeological
    Research
Downing Street
Cambridge CB2 3ER
UK
jsp1005@hermes.cam.ac.uk

MARGARET SERPICO
Institute of Archaeology
University College London
31-34 Gordon Square
London WC1H 0PY
UK
mt_serpico@hotmail.com

IAN SHAW
Department of Archaeology (SACOS)
University of Liverpool
14 Abercromby Square
Liverpool L69 3BX
UK
imeshaw@supanet.com

ANDREW SHORTLAND
Research Laboratory for Archaeology
    and the History of Art
6 Keble Road
Oxford OX1 3QJ
UK
andrew.shortland@archaeology-
research.oxford.ac.uk

LAURENCE SMITH
McDonald Institute for Archaeological
    Research
Downing Street
Cambridge CB2 3ER
UK
ls101@cus.cam.ac.uk

RACHAEL THYRZA SPARKS
Pitt Rivers Museum
60 Banbury Road
Oxford OX2 6PN
UK
rachael.sparks@prm.ox.ac.uk

MICHAEL TITE
Research Laboratory for Archaeology
    and the History of Art
6 Keble Road
Oxford OX1 3QJ
UK
michael.tite@archaeology-
research.oxford.ac.uk

Preface and Acknowledgements

In September 2002, a second workshop on the theme of the social context of technological change was held at the McDonald Institute for Archaeological Research, University of Cambridge. It followed a meeting in Oxford, two year's earlier which resulted in a book, *The Social Context of Technological Change. Egypt and the Near East, 1650–1150 BC*, edited by Andrew Shortland and published in 2001 by Oxbow Books. The format of both meetings was the same: each paper was followed by 20 minutes of discussion led by a participant who had read the paper in advance. The format is not yet a common one but it has the great merit of ensuring an informed discussion and helping to overcome the problem for an audience in absorbing complicated data for the first time and having enough energy left to ask sensible questions.

Discussion has been the core of these meetings so far since their aim is to relate the results of the specialist investigator to broad historical questions concerning the nature and development of ancient societies. Paradoxically, the spur for these meetings has come as much from two recent publications as from discussion: P. R. S. Moorey's, *Ancient Mesopotamian Materials and Industries: the Archaeological Evidence*, Oxford, 1994, 1999 and P. T. Nicholson and I. Shaw (eds), *Ancient Egyptian Materials and Technology*, Cambridge, 1999, both evolving from Alfred Lucas, *Ancient Egyptian Materials and Industries*, London 1926 (1st edition); London 1962 (4th edition expanded by J.R.Harris). Both publications illustrate the richness of textual, archaeological and scientific evidence which confronts students of the technologies of these great civilisations. Equally, they demonstrate the complexity of the sources; the difficulty of their interpretation and of the integration of the different kinds of evidence they provide. A good example of this is the source material for the study of glass-making (Moorey 1999,189–215; Nicholson and Henderson in Nicholson and Shaw (eds) 1999, 195–224; and, in Andrew Shortland's volume: Robson; Shortland, Nicholson and Jackson; Shortland; and Rehren, Pusch and Herold. To make matters worse scholars have tended to confine themselves to a particular methodology and type of evidence and traditionally they have worked alone. This approach is appropriate for some investigations but not for all, as the multi-authorship of some of the papers indicate. However, the scientific research group model, in which a group of specialists come together, under direction, each contributing to the overall research design as well as to its performance, is still rare in archaeology.

For the Cambridge meeting it was decided to enlarge the discussion geographically to include the Aegean formally (several Aegean related papers had in fact appeared in the Shortland volume) and thematically to invite papers on natural products and raw materials. The time frame was not changed since it was felt that there was still much to explore in the period of the Late Bronze Age/ New Kingdom. As a collection of papers there are some differences between this volume and the first. A majority of the papers draw on Egyptian evidence but this should be taken as a result of the stimulus of the wealth of Egyptian sources rather than a reflection of the editors' interests. Moreover, two papers were given at the meeting, by Sariel Shalev, "Metal production in the beginning of Iron Age Israel: facts and fictions", and by Neil Brodie and Ian Whitbread, "Technological Traditions and Ceramic Exchange in Laconia, Greece during the Middle Bronze Age", which could not be included in the publication and they dealt respectively with the Aegean and the Levant.

The papers published here illustrate a multiplicity of approaches to the problems set by ancient technologies, modelling (Shortland); methodology of art history and archaeology applied to a problematic group of artefacts (Ashton; Phillips); integration of archaeological and textual sources for a review of basic questions of identity and status (Shaw; Sparks); and the application of the results scientific analysis with archaeological evidence to illuminate technology (Panagiotaki *et al.*; Martlew; Serpico; Smith *et al.*) or a supposed innovation (Bourriau).

The editors would like to thank Laurence Smith for his help in the planning and execution of the arrangements for the meeting itself, Dr. Chris Scarre, Professor Lord Renfrew and the Managing Committee of the McDonald Institute for use of the facilities of the Institute and Dora Kemp of the McDonald Institute for her help with the technical production of the volume.

Chapter 1

Hopeful Monsters? Invention and Innovation in the Archaeological Record

Andrew J. Shortland

Abstract

The processes of innovation are discussed on a theoretical level drawing on archaeological and historical examples. The use of evolutionary theory, particularly the theory of 'hopeful monsters,' is examined and its applicability to cultural systems assessed. Three stages of innovation are proposed: discovery, invention and innovation/diffusion, each having different controlling factors. The innovation of glassmaking is then considered in more detail, looking at the technological, social and cultural factors that might have affected the adoption of glass as a new technology in the 15th century B.C. The overwhelming influence of cultural over technological factors in the success of glass technology is proposed.

INTRODUCTION

As would be expected, many technologies can be seen to have a short period of time where they go from 'unknown' to 'widespread' in the archaeological record, i.e., an 'innovation' occurs. These innovative bursts often appear to be almost instantaneous and have been used to mark the beginning of archaeological epochs and name the time periods that follow, giving us the Bronze and Iron Age, for example. However, in many technologies (e.g., bronze, glass, iron), rare and apparently random early occurrences of the material have been found before the main innovative explosion. This paper considers what these rare finds reveal about the mechanisms of invention and innovation and how this affects our understanding of technological change in general. It goes on to consider in more detail what and how geographical, technological and cultural factors affect inventive and innovative rates. It derives examples particularly from the first production of glass and goes on to apply the theory to the discussion of how glassmaking might have been discovered.

The process of technological change is:

Stage 1	Stage 2	Stage 3
DISCOVERY	INVENTION	INNOVATION
The find	Application	Diffusion

Technological change is a ubiquitous feature of the archaeological record and represents one of the most important tools for the dating of archaeological horizons available to archaeologists. However, the actual process whereby these changes come about are not often considered in detail. In order to examine it, this paper divides the process into three stages, each of which is controlled by different factors. The first two are loosely based on Peltenburg's stages of glass production (Peltenburg 1987), discussed at greater length below. The first stage is called in this paper the 'discovery' stage. This is the stage that creates, for the first time, a novel material or process. It moves from the raw materials and applies existing or novel techniques and finishes with a product that has, as yet, no use.

The second stage of the process is closely related to the first. This is the 'invention' stage, and starts with the product derived from the discovery. The first part of the innovation is the realisation that the novel product of invention has a potential use, and a potential market to use more modern capitalist terminology. This 'application' of the invention is then worked up into a fully working finished product.

The third part of process is the 'innovation' stage, the diffusion of the invention throughout the populace as a whole and then potentially throughout neighbouring societies as well. Following the terminology used in this paper, this diffusion marks the change of the invention into a true innovation.

It is important to realise that the discovery, invention and innovation stages are often separate not just in the model described above, but in reality as well. A significant number of products lie 'dormant' after the discovery stage and before one or both of the invention and innovation stages. The example usually quoted for this is the fax machine. The patent for the fax machine was granted on 27 May 1843 to Alexander Bain (*Encyclopaedia Britannica*). This was some 33 years before Alexander Graham Bell filed his patent for the telephone (14 February 1876). However, the first, commercial fax service was not opened until 1865, when a very limited link between Paris and Lyon was set up. Faxes only really came into their own in 1906 when they found their first major use, to transmit photos for newspapers. Thus between discovery and innovation, a period of 63 years elapsed.

STAGES 1 AND 2: MODES OF DISCOVERY AND INVENTION

The two different ways in which discovery might come about are accidentally and intentionally. These two opposites appear to fall as end members on a spectrum of different possibilities, actually a continuous series involving more or less accident or intent. Accidental discoveries or inventions are the stuff of scientific folklore. One of the best examples of this is penicillin and is discussed below. It is interesting to note that Pliny describes the first production of glass as an accident:

> "A ship belonging to traders in soda once called [at a beach at the mouth of the Belus River], so the story goes, and they spread out along the shore to make a meal. There were no stones to support their cooking pots, so they placed lumps of soda from their ships under them. When these became hot and fused with the sand on the beach, streams of an unknown translucent liquid flowed, and this was the origin of glass." *(Pliny the Elder 1991, Book XXXVI.191)*

This aetiological story, if it were true, would be right at the accidental discovery end of the spectrum of possibilities. Within this class of accidental discoveries, two different forms can be discerned, although they are not necessarily exclusive. The first is where the accident derives from a unintentional mistake made in another manufacturing process: perhaps the wrong raw material is added, or it is heated to long or too strongly. The second is more similar to the glass origins story of Pliny, that is to say where a random observation of an event unconnected with a manufacturing process leads on to a discovery.

The other end of the spectrum is deliberately going out to discover new materials or techniques. This is very much the modern idea and can be divided once more into two groups. The first is material- or method-led, and involves searching for new applications for a product or production technique already available for other things. Modern drugs research tends to be very much along these lines. Here, a group of new compounds is produced by making small structural changes in existing drugs and the novel compounds tested for activity against a series of pathological problems. The production of the compounds and their testing tends to be very systematic, and this approach can yield good results. This is the 'experimental' approach of hypothesis-test-result-conclusion. The second way of deliberately attempting to create discoveries is to follow the same type of methodology, but from the opposite direction, that is to say to start from the problem and try to find a suitable product that will fill that role. This could be termed the 'inventive' approach, and is the methodology followed by many modern small scale inventors.

Between the accidental and the deliberate there are a whole range of possibilities with elements of both. Just one might be picked out and that is the possibility of reuse of a by-product. Many ancient technologies, especially those involved with metallurgy, can produce a whole range of by-products during their processing. It is possible that some of these by-products, deliberately produced as a consequence of the smelting of a particular metal, for example, may have been recognised as possibly having further uses in other processes. For example, Jennifer Mass (Mass *et al.* 2002) has suggested that litharge from silver smelting might have been subsequently used for the production of pigments in glass making. In this case the 'discovery; stage happens often, in that the novel object is produced over and over again in the production of another material. However, the invention stage – converting the object into something useful – is delayed.

To illustrate the inventive process and a number of other aspects of discovery, invention and innovation, a modern example, that of the discovery of penicillin, will be considered in more detail.

PENICILLIN AND FLEMING: 1928

The discovery of penicillin is one of the great stories in the history of science (Macfarlane 1985). It is one of those stories the details of which are frequently garbled and the outcome somewhat different to the version most commonly heard. Alexander Fleming was born in Scotland in 1881, read medicine and in the 1920s was researching antiseptics. He was by his own admission not the tidiest of workers and one day in September 1928 he had left a Petri dish open on his bench while he went on holiday. It was already loaded with staphylococci, and spores of a fungal mould (*Penicillium notatum*) floated in through an open window and settled on it. On his return Fleming found that the bacteria on the Petri dish had not grown as they normally would have done and deduced that they had been killed by something produced by the mould, which he called penicillin.

This is usually where the story ends, with the fairytale discovery of the first antibiotic. But this was not the full story and it is interesting to note for this paper what really happened. Fleming published his results, but a combination of minor technical difficulties reproducing results, its lack of effect against certain bacteria (e.g. cholera, bubonic plague), and other pressing research meant that Fleming's interest waned. Penicillin therefore faded from the scientific scene and no further work was attempted for over ten years.

The research was continued in Oxford by Howard Florey, Ernst Chain and Norman Heatley, who were working on microbial antagonists. Chain found Fleming's original report, and they succeeded in isolating the active ingredient from the liquid produced by the mould and produce enough antibiotic to test its worth. On 25 May 1940, less than a year after starting the work, they inoculated eight mice with a lethal dose of streptococci and then injected four of them with penicillin. Next day the four mice given streptococci alone were dead, the four with penicillin were healthy. Subsequent successful tests were carried out on humans and full scale production began. In 1945, Fleming, Chain and Florey were awarded the Nobel prize for their work.

It is interesting to note that, while there was a technical step to be overcome in the process of producing penicillin, the driving force behind this was the huge rise in demand for such a drug brought about by the Second World War. As Clara Solomon wrote in 1861 at the beginning of another war (the American Civil War), "necessity and war is [sic] the mother of invention" (Ashkenazi 1995).

STAGE 3: INNOVATION AND DIFFUSION – DOI THEORY

Much work has been devoted to innovations and the processes that control their spread. This is hardly surprising, given the commercial importance of the subject.

However, in terms of this paper it is important to note that almost all the work has been limited to examining the last of the three stages defined above, that is to say the diffusion of the innovation. This work has spawned a whole field of study, known as diffusion of innovation (DOI) theory. One of the most important contributions to the development of this theory was Everett Rogers' *Diffusion of Innovations*, first published in 1962 and now in its third edition (Rogers 1983).

While various terms have been used, five stages through which an innovation must pass during its diffusion have been recognised. The first is *knowledge* of its existence, and understanding of its functions, followed by *persuasion*, which is the forming of a favourable attitude towards the innovation. Thirdly a *decision* has to be made, that is to say a commitment to its adoption and then its *implementation*, the putting of the innovation to use. Finally comes the *confirmation* stage, which revolves around a reinforcement based on positive outcomes deriving from the innovation. How fast the innovation passes through these stages is based on the characteristics of an innovation. Once again, five features of innovations have been identified that affect the rate of adoption. These are *relative advantage*, which is the degree to which the innovation is perceived to be better than what it supersedes, its *compatibility* with existing values, past experiences and needs. High *complexity* makes it difficult to understand and use, has a negative affect on the rate, whereas high trialability (the degree to which it can be experimented with on a limited basis) and high *observability* in the results of the trials has positive affects.

DOI Theory also recognizes that the rate of innovative change is not constant throughout a population. Different adopter categories have therefore been proposed to describe the actions of the different types of respondants. These range from *innovators*, who are leading the innovation and represent those most easily swayed by the change, through *early adopters* and the *early majority* to the *late majority* and finally the *laggards*. Other authors have added *reactionary*, *curmudgeon* and *iconoclast* classes to express yet slower adopters. Within the adopters, certain groups or individuals can also have specific roles within the diffusion of the innovation. One of the most important of these is the *change agents* who positively influence innovation decisions, by mediating between the proponents of change and the relevant social system. These may be assisted by *change aides* who are more embedded in the culture and therefore closer to the target adopter group. To further complicate matters, it has long been recognised that different individuals act in different adopter roles with different innovations, making for an extremely difficult predictive problem.

However, a general pattern can often be seen with many innovations and this is often expressed in the innovation adoption curve (Figure 1.1). This curve predicts that rate of adoption will be slow initially and latterly. The initial slowness is due to the need for a critical mass of adopters to be formed. After this critical mass has been achieved, the innovation tends to spread itself by chain reaction, the beginning of which is defined as the *take off* point. Eventually the innovation fills the market niche available for it and saturation is achieved, slowing down the rate of change again. This curve allows the definition of the

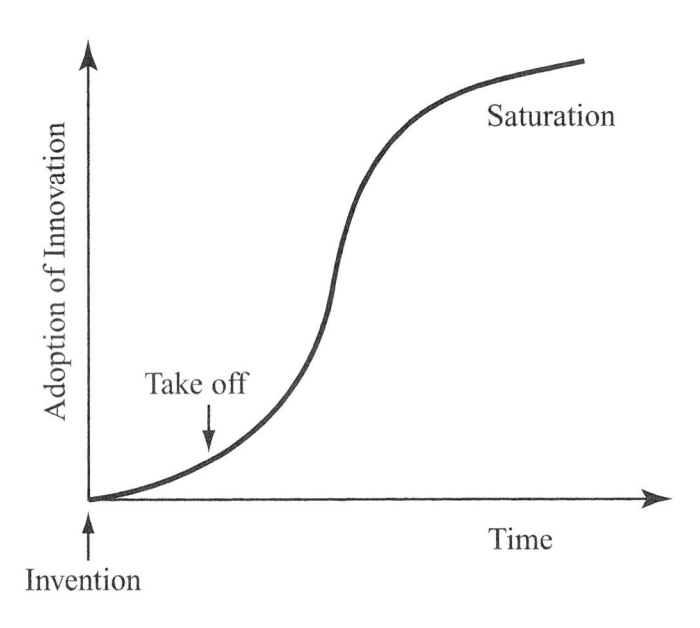

Figure 1.1 Innovation adoption curve.

two stages of innovation to be clearly seen. The application stage of innovation occurs before the take off point and the diffusion of the innovation after that critical point.

DOI Theory, while being interesting in so far as it goes, has a number of limitations. Firstly, and most importantly for this paper, it deals only with the final stage of the discovery/invention/innovation process, whereas the interest to the student of ancient technology is at least as much in the early stages. Secondly, even within that final stage, DOI Theory is more descriptive than explanatory in that it does not attempt to explain why certain individuals behave in certain ways, merely stating that they do and exploring the mathematics of the consequences of that statement. It is also particularly hard to apply the theory to ancient cultures where individuals and groups are difficult to identify and motives obscure. Even where it is possible to do so, its low explanatory power is not helpful in resolving the 'whys' and 'hows' of ancient technological change. It is therefore necessary to go back to the beginning and examine the invention/ discovery and application/innovation in more detail.

SOCIETY AS ENVIRONMENT – EVOLUTION AND INVENTION

One method that has been used to examine invention in the archaeological record is by analogy to biological evolutionary theory. The use of evolutionary theory to model the development of cultural systems has received a great deal of attention over the last fifty years or so. Dunnell considers that modern evolutionary biology provides an explanatory framework for the processes of cultural change, but cautions that it cannot "be applied un-emended and uncritically to cultural phenomena, be they ethnographic or archaeological" (Dunnell 1981, 37). Dunnell considers that the processes of natural selection, gene flow, gene drift, and mutation all have analogous processes in both biological and cultural evolution. When evolutionary theory has been applied to invention/innovations, it has usually been attached to the diffusion modelling of the innovation itself. However, certain theories can also be used to examine invention as well. The one used here is the idea of 'hopeful monsters.'

All genetic systems are liable to error when they reproduce. These mutations cause variation in asexual reproduction and add to the variation that is already built into sexual reproduction. They have been used to date speciation events, for example in the 'Out of Africa' hypothesis and 'Mitochondrial Eve' (Cann *et al.* 1987). However, in the 1930s Richard Goldschmidt suggested that the explanation might rather lie in what are known as embryological monsters, such as the occasional birth of a two-legged cow or a three eared mouse (Goldschmidt 1940). Although he admitted that such monsters rarely survived for long, yet given geological time, some monsters might actually be better suited to survive and reproduce than their normal siblings. These extremely unusual successes might cause some speciations without intermediate stages. Goldschmidt named this idea the "hopeful monster theory" and it caused an uproar in the scientific community, being dismissed with derision. There it lay dormant until the work of Eldridge and Gould in the 1970s, who were similarly concerned about the lack of evidence in the fossil record for the gradualism of Darwinian evolution. They therefore proposed "punctuated equilibria" – periods of relative stasis with evolutionary bursts in between (Gould and Eldridge 1972, 1977). They also revived the 'hopeful monster' theory, close related to punctuated equilibria. Analogy between biological theories to invention and innovation can be drawn. Regularly within the practises of ancient industries, accidents would have happened and mistakes in processing have been made. These would usually have been thrown away as wasters, or perhaps fashioned into small low-status objects, but with no attempt made to repeat the 'mistake.' Therefore, throughout technological history there is the chance of small 'wasters' in other industries dealing with similar raw materials or techniques presaging a new technology. However, obviously no conscious link existed in the minds of those producing it that it might lead to a future innovation. There is a similarity therefore, to the hopeful monsters of evolutionary theory – dead ends with no future in themselves. However, very rarely, one of these technological wasters might strike someone as a not so much a mistake as a new possibility. If an application can be found, it becomes an invention and if it proves popular, the innovation can occur. The technological 'hopeful monster' turns into a successful innovation.

There are however, strong criticisms that can be drawn to applying biological hypotheses to cultural events. Wenke and others have completely rejected the use of evolutionary theory as an analogy in archaeology. He considers that there are four main reasons for this rejection (Wenke 1981, 111). Firstly, there are very different systems of character transmission in the two systems, the cultural one relying on a complex web or tradition, history and connectivity that functions very differently in different temporal or geographical areas. A straight analogy is therefore much too simplistic. Secondly, cultural processes operate on a much shorter time scale than geological ones. This might be especially true when considering biological 'hopeful monsters,' where extremely long timeframes are necessary. Thirdly, there is an absence of testable concepts of natural selection in the cultural sphere. Finally, Wenke cites a suspicion that cultural phenomena are different from all others and can usefully be examined only within cultural terms. There is much in what Wenke says, and the uncritical use of evolutionary theory analogies has been greatly unhelpful. Yet it is important to remember that behind biological evolutionary theory is a mathematical model that will work in any circumstance where there is variation and selection. The mechanisms that give rise to variation and selection may be hugely different in different spheres, but the underlying mathematical model is the same. There is therefore no *prima facie* reason why the models should not be used. The possibility therefore remains that even though they are very different, analogies between evolution and innovation can create interesting debating points for use in examining cultural phenomena and, if treated as the models and suggestions they are, have the potential to be useful.

THE INNOVATION OF GLASSMAKING

Glass was first produced on a large scale around 1500 B.C. in both Mesopotamia and Egypt. It is probable that the innovation originated with an invention in northern Syria (Nicholson 1993), and both the material and technology diffused rapidly from there. By the Amarna period (the mid-14th century B.C.), glass was reasonably common, although still a prestige, high-value commodity. There was certainly a demand for the material, with a number of factories present at Amarna (Nicholson 1995, Petrie 1894, Shortland 2000a) and the near contemporary Malkata (Keller 1983). There is also evidence from the Amarna letters that the Egyptian king at this time was requesting glass to be sent from Syria in the form of tribute (Moran 1992). Glass was used to produce beads, inlays, figurines (especially in Mesopotamia) and its signature object, the core-formed glass vessel.

However, there are scattered finds of glass dating to before 1500 B.C. throughout Egypt and the Near East. Peltenburg (1987) lists these finds, and although a number of them have doubtful provenance and have been questioned (Lilyquist and Brill 1993), a body of early finds that are undoubtedly 'real' remains. Peltenburg defines these finds as glass of "Stage 1, a pre-adaptive stage characterised by the very infrequent and irregular use of glass." This corresponds to Stage 1 of the scheme outlined above. Peltenburg Stage 2, is when the "individual properties [of glass] have first been realised" (Peltenburg 1987, 18), which is the equivalent again of Stage 2 in the model above. Stage 3 (absent in the Peltenburg model) then reflects the spread of this glass away from its place of first manufacture to Egypt and the Aegean. Peltenburg draws on the use of iron, where (as discussed below), a similar pattern is found.

Various suggestions have been made why glass was sporadically produced through Stage 1, but this innovation (Stage 3) did not occur until around 1500 B.C., some 1000 years after the first glass appeared. The presence of the glass shows that there is no technological reason why glass-making could not have started earlier, indeed on a small scale it obviously did. However, it might be possible to point to technical developments in the ability to work glass in its hot state as the key stage in the triggering of the development of Stage 3. Specifically, it has been proposed that core-forming was the major development within the necessary glass-working skills (Peltenburg 1987).

Core-forming and its associated trailing and marvering are the skills required for the production of glass vessels. However, not all of the earliest glass vessels are core formed. Some appear to have been cold worked, and some may have been formed in a mould (Shortland, in press). Indeed, it seems in Egypt at least that the first use of glass may have been as a semi-precious stone by the existing and thriving stone working industries. So core-forming and other hot working aspects of glass-working were not necessary in order to produce glass vessels or to work with glass.

There is also the question of how important glass vessels were, in terms of their volume within the production of glass in Stage 3. Glass vessels, perhaps due to their polychrome nature and striking designs, are very obvious products of the glass industry in the 2nd millennium. However, maybe simply because of their striking nature, the percentage of the total volume of glass produced that actually went into these vessels may have been exaggerated. It is difficult to know what percentage of the glass went into vessels and what into beads, amulets and inlays. A survey of glass finds over part of the site of Amarna (Shortland 2000b) showed that of the 113 glass objects found by the German excavations of the early 20th century, only 19 were fragments of glass vessels. It seems probable, therefore, that significant amounts of the Stage 3 glass is being used in non-vessel products. Once again, this tends to suggest that core-forming was not the step that led to the innovation and diffusion of glass. It therefore seems probable that there are no technological reasons connected with glass-making or glass-working that prevented the movement of glass beyond Stage 1. The 'rate-determining steps' therefore seem to be in the less technological and more socio-cultural Stages 2 and 3.

In looking for a reason for glass innovation, Peltenburg turns to the similar iron industry. Here too the sporadic production of iron (Stage 1) is succeeded by rapid innovation and spread (Stages 2 and 3). Peltenburg sees a "consensus view that some historical element is responsible for the adoption of iron as a working metal." He particularly points towards perceived shortages in copper and tin for the manufacture of bronze tools that appear to correspond with iron innovation. Iron tools therefore step in where bronze tools are difficult to obtain. Peltenburg does not see the same thing happening in glass; he states "it seems hardly likely that the crucial factor for [glass] adoption is a severe shortage of rare raw materials" (Peltenburg 1987, 18). However, it could be argued that this is exactly the case. Glass is seen as a substitute for rare semi-precious stones such as lapis lazuli and turquoise. In the 15th century B.C., conspicuous display and competitive gift-giving became an important feature of states in the Near East (Kemp 1989), almost certainly causing the demand for such stones to rise. This would have strained supply lines for these minerals and perhaps caused palace workshops to look around for viable alternatives among the existing crafts. Glass adoption could therefore have been partly due to "a severe shortage of raw materials". However, Peltenburg goes on to insist that "the pre-disposition of society... is the essential factor in the dynamics of invention" (Peltenburg 1987, 18). This certainly seems to have been the case. Put into other words, it is the 'environment' into which the discovery is introduced that determines whether it succeeds in passing through to the invention and innovation stages. In the case of glass, and possibly also for other technologies that seem to go through a long run up of 'hopeful monster' stages, the cultural environment into which it was introduced needed to be correct.

CONCLUSIONS

Evolutionary models have only a limited utility in examining technological change in the archaeological record. However, if applied in the right way, they can provide thought-provoking analogies that enable technological processes to be explored. The idea of 'hopeful monsters' as an analogy to describe the rare finds of early 'Stage 1' glass is just such as case. Such rare glass objects seem to represent errors in the manufacture of perhaps faience and metals, showing that there was no technological reason why glass could not have been produced earlier that 1550 B.C. However, after that date it appears that the socio-cultural 'environment' changed so as to allow the glass innovation to progress to Stages 2 and then rapidly to Stage 3, ending with the diffusion of glass throughout the Near East.

ACKNOWLEDGEMENTS

The author would like to thank Dr Alexander Wagner (North Dakota State University) for many helpful and fruitful conversations on this topic over the years.

REFERENCES

Ashkenazi, E. (ed) 1995, *The Civil War diary of Clara Solomon: growing up in New Orleans, 1861–1862,* Louisiana State University Press, Baton Rouge.

Cann, R. L., M. Stoneking and A. C. Wilson, 1987, Mitochondrial DNA and Human Evolution, *Nature,* January 1, 1987, 31–36.
Dunnell, R. C., 1981, Evolutionary Theory and Archaeology, in Schiffer, M. B. (ed.), *Advances in Archaeological Method and Theory: Selections for Students from Volumes 1 through 4.* , 35–90. Academic Press, New York.
Goldschmidt, R., 1940, *The Material Basis of Evolution,* Yale University Press, New Haven.
Gould, S. J. and N. Eldridge, 1972, Punctuated Equilibria: An Alternative to Phyletic Gradualism, in T. J. M. (ed), *Models in Paleobiology,* Shopf, 82–115. Freeman, Cooper and Co., San Francisco.
Gould, S. J. and N. Eldridge, 1977, Punctuated Equilibria: The Tempo and Mode of Evolution Reconsidered, *Paleobiology,* 3, 115–51.
Keller, C. A., 1983, Problems of dating glass industries of the Egyptian New Kingdom: Examples from Malkata and Lisht, *Journal of Glass Studies,* 25, 19–28.
Kemp, B. J., 1989, *Ancient Egypt: Anatomy of a civilization,* Routledge, London.
Lilyquist, C. and R. H. Brill, 1993, *Studies in Ancient Egyptian Glass,* Metropolitan Museum of Art, New York.
Macfarlane, G., 1985, *Alexander Fleming: The man and the myth,* Oxford Paperbacks, Oxford.
Mass, J. L., M. T. Wypyski and R. E. Stone, 2002, Malkata and Lisht glassmaking technologies: towards a specific link between second millennium BC metallurgists and glassmakers, *Archaeometry,* 44 (1), 67–82.
Moran, W. L., 1992, *The Amarna Letters,* The Johns Hopkins Press, Baltimore.
Nicholson, P. T., 1993, *Ancient Egyptian Faience and Glass,* Shire Egyptology, London.
Nicholson, P. T., 1995, Recent excavations at an ancient Egyptian glassworks: Tell el-Amarna 1993, *Glass Technology,* 36 (4), 125–128.
Peltenburg, E. J., 1987, Early faience: recent studies, origins and relations with glass, in Bimson, M. and I. C. Freestone (eds), *Early vitreous materials,* British Museum Occasional Papers, 56, 5–30. London.
Petrie, W. M. F., 1894, *Tell el-Amarna,* London.
Rogers, E. M., 1983, *Diffusion of Innovations,* The Free Press, New York.
Shortland, A. J., 2000a, The number, extent and distribution of the vitreous materials workshops at Amarna, *Oxford Journal of Archaeology,* 19 (2), 115–34.
Shortland, A. J., 2000b, *Vitreous materials at Amarna: the production of glass and faience in 18th dynasty Egypt,* British Archaeological Reports International Series, S827, Archaeopress, Oxford.
Wenke, R. J., 1981, Explaining the Evolution of Cultural Complexity, in Schiffer, M. B. (ed), *Advances in Archaeological Method and Theory: Selections for Students from Volumes 1 through 4.,* 79–127. Academic Press, New York.

Chapter 2

Identity and Occupation: how did individuals define themselves and their work in the Egyptian New Kingdom?

Ian Shaw

Abstract

What trades or professions were recognised by the ancient Egyptians during the New Kingdom, and how can we hope to identify them either textually or archaeologically? The Egyptian reliefs and paintings showing certain types of skilled work in progress are dominated by particular types of craftwork, such as agriculture, carpentry, stone-working and metalworking, providing far less information about the production of glass or faience, the building of houses, quarrying of stone, and numerous other activities that must in reality have been of equal significance outside the funerary context. The domination of Egyptian cultural material by a combination of the influence of scribal/bureaucratic ideals and the desire to be seen to fulfill certain roles in the afterlife tends to prevent us from fully recognising the ancient Egyptians' day-to-day consciousness of their roles as particular tradesmen or professionals. This paper therefore looks at a combination of different types of evidence from the New Kingdom, including a detailed archaeological case-study at Amarna, and considers possible approaches which may allow us to gain a better understanding of the Egyptians' 'occupational identities.'

INTRODUCTION

The origins of this paper are partly to be found in a pair of late 18th Dynasty letters published more than 70 years ago by Eric Peet (1930a). Although these two documents (*P. Robert Mond 1 & 2*) were found in the tomb of Hes at Thebes, they are in effect the only two surviving texts on papyri originating from Amarna since they were written by a resident of Akhetaten, called Ramose. The contents of the letters are not especially illuminating (see Peet 1930a), except for the fact that Ramose describes himself as an 'oil boiler in the estate of princess Meretaten' (or 'unguent preparatory,' according to Wente 1990, 94–6). Two questions that immediately arise are: where (and how) would an oil boiler have worked, and, more importantly perhaps, how would we recognise a worker of this type in the archaeological record?

At best, such trades as oil boiler, washerman or incense roaster might have left material remains which were diagnostic of their particular trade but not necessarily recognisable as such by archaeologists. At worst they might have characteristically left nothing in the way of distinctive artefacts within their houses or courtyards. It is likely that many of the houses at New Kingdom urban sites such as Amarna might have been occupied by specialised workers whose trades or professions left much more subtle traces than the crucibles of the coppersmith, the stonemason's chisel or the carpenter's adze.

FUNERARY EVIDENCE FOR CRAFTS AND PROFESSIONS

One of the earliest systematic attempts to examine the evidence for craft specialisation in Egypt at the beginning of the historical period deals with artefactual evidence from the Predynastic and Early Dynastic phases, when textual evidence is of course relatively sparse. Whitney Davis (1983) studied the range of artefacts in Predynastic tombs at Naqada in order to identify skilled artisans by the special tools of their trade. He concludes that "most crafts were a household enterprise, with only the more difficult crafts of ivory-working and stone-working taken up by individual specialists, but the homogeneity of the tombs throughout the cemetery suggests that skilled workers might have been economically and socially indistinguishable from the rest of the community" (Davis 1983, 127).

However, Davis' examination of Early Dynastic tombs at such sites as Saqqara and Abydos indicates that the uniformity of the Predynastic community was being gradually transformed during the crucial transition to statehood. He suggests that the subsidiary tombs surrounding those of rulers such as Djer and Djet, at Abydos, represent evidence of the appearance of a class of professional artisans, clearly differentiated from the rest of population: "the appearance of the royal serekh on tools deposited in Early Dynastic subsidiary burials could be interpreted variously, but whether we take the objects to be gifts of the monarchy or simply signs of professional affiliation, a close relationship between royal patron and craftsmen must be assumed" (Davis 1983, 132). He also argues that a private tomb such as M13 at Abydos, which contained an adze decorated with the royal serekh, provides evidence for the appearance of wealthy artisans serving the 'community' rather than working specifically for the ruling élite.

Just as the graves examined by Davis ccasionally seem to contain items key to the occupation of the deceased, so some early private statues show the individual holding the essential tool of his trade. The Early Dynastic statuette of Ankhwa (or Bedjmes) the boat-builder, in the British Museum, is one of the classic examples of this genre. By the Old Kingdom, however, the bureaucratic and hierarchical system of government had apparently developed to such an extent that skilled craftsmen were rarely having themselves explicitly portrayed with the tools of their trade (or indeed including such implements in their funerary equipment). Skilled workers continued to be represented but in the form of so-called 'servant statuettes' and scenes of craftwork on the walls of the tomb. Craftsmen and other skilled workers had in a sense become part of the background in the funerary context, rather than holding centre stage. The emergence of the scribe as the predominant human element of the Egyptian economy seems to have had the effect of eclipsing most other professions as far as the identities of the deceased were concerned.

There were many early Old Kingdom important officials who held a variety of posts in which they were in charge of skilled craftsmen or practising a specific profession, but they seem to have generally chosen to accentuate their role as scribe. Thus, as early as the 3rd Dynasty, Hesyra, whose titles show him to have been a dentist and physician, simply represents himself holding the pen, ink and palette of the scribe. This scribal domination continued through to the New Kingdom, when we find that a chief of artists such as Bak represents himself with no visual indication of his profession.

As well as the domination of the scribal archetype, there is also a tendency for the surviving visual evidence to emphasise certain types of work at the expense of others. Numerous reliefs and paintings throughout the pharaonic period show skilled work in progress, but these are dominated by particular types of craftwork, such as agriculture, carpentry, stone-working and metal-working, making far fewer references to such activities as the excavation of tombs, the building of houses, quarrying of stone, the production of glass or faience, and numerous other types of work that must in reality have been of equal significance outside the funerary context. There are occasionally interesting exceptions, such as the scene of chariot repair at Qadesh and the chariot repair shop shown in a tomb of the New Kingdom at Saqqara, but usually the funerary repertoire of crafts and professions is very repetitive and restricted. As Chris Eyre (1995, 55) points out, "the pig-eating fish-eating world of the Egyptian town hardly appears in the beef-eating poultry-eating world of the temple and tomb."

When individuals are depicted at work in New Kingdom tombs and temples it is obviously essential to bear in mind the specialised context and the system of decorum involved. Above all there is the symbolic role of funerary art, in which, for instance, both the tomb-worker Sennedjem and the generalissimo (and future king) Horemheb show themselves working in the fields in the afterlife, in order to play their part in the scenario laid down in the *Book of the Dead*, whereby Spell 110 requires the deceased to plough and reap: "I plough and I reap and I am content in the City of God. I know the names of the districts, towns and waterways which are in the Field of Offerings and of those who are in them; I am strong in them and I am a spirit in them."

Identity and Occupation in New Kingdom Egypt 15

ARCHAEOLOGICAL FACTORS THAT OBSCURE EVIDENCE OF CRAFT SPECIALISATION

Since the visual evidence of work practices during the dynastic period is very selective and restricted, a greater onus is placed on settlement archaeologists to try to address this problem. The artefactual material from Egyptian town sites arguably constitutes an account of human behaviour that is somewhat less 'artificial' or 'contrived' than that from the cemeteries. A bronze adze found in a tomb, for instance, may be anything from a religious symbol to an expression of wealth or power (particularly if it bears a potent symbol, such as the royal serekh in the Early Dynastic tomb M13 at Abydos, mentioned above), but an adze found lying on the floor of one of the rooms or outbuildings of a house is clearly much more likely to represent evidence of the actual practice of carpentry. The architectural and artefactual surroundings of an adze found in a house can also sometimes provide a much richer context within which the behaviour and status of the tool's owner may be assessed.

Although it is possible to gain a superficial indication of the range and relative importance of activities across the city of Amarna as a whole (see the barchart in Figure 2.1), this kind of quantitative information can obviously be biased by many factors of preservation and archaeological visibility. It is not simply that

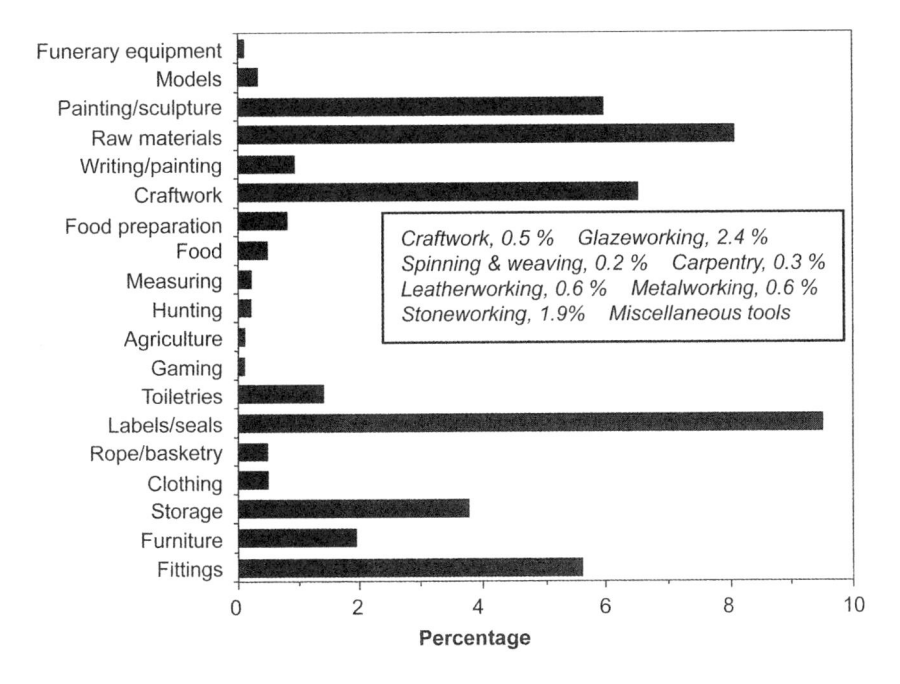

Figure 2.1 Barchart showing the proportions of different types of human activity represented in the artefactual record across the whole city at Amarna.

certain activities may be altogether lost from the archaeological record but that some may, by their nature, leave more or less debris than others. It is therefore difficult to infer the relative importance of activities from barcharts recording quantities of objects, since the charts cannot be automatically 'converted' into 'hours of work' or 'numbers of craftworkers.' A higher percentage of spinning and weaving artefacts in the North Suburb than in the North City can be fairly securely interpreted as an indication of a socio-economic difference between these two sectors of the Amarna community. However, a higher percentage of carpentry artefacts than leatherwork artefacts within the North City may indicate not so much more man-hours of carpentry as the simple fact that carpentry has left more traces than leatherwork in the archaeological record.

The degree of preservation of different types of material remains can vary immensely even within a single Egyptian city such as the one at Amarna. Thus, for instance, the fact that the Amarna 'workmen's village' is situated at a greater distance from the Nile, compared with the main city at Amarna, seems to have ensured that wood, matting and textiles were much better preserved in the village than in the city. When it is further considered that some crafts actually require more tools and leave greater amounts of débitage than others, then the inadequacy and complexity of the data become even more evident. The flow chart in Figure 2.2 shows how a typical set of 18th Dynasty carpentry tools may be gradually diminished, in the archaeological record, by the various factors of preservation and inadequate excavation.

PATTERNS OF CRAFTWORK AND PROFESSIONS AT AMARNA

Taking Amarna as our most 'data-rich' case-study for such patterns of craftwork in New Kingdom towns, it is useful not only to ascertain the importance of a particular occupation relative to others practised within a single residential zone, but also to see how this relative importance varies from one zone to another. It is, for instance, interesting to note that tools associated with craft-production constitute 31% of the total artefacts in the 'South Suburb' at Amarna, but it is considerably more significant to know that this percentage is much higher than that in the 'workmen's village' (24.1%) or the 'North City' (23.9%). The numbers of craft tools from a single suburb or even a single house can only be properly assessed when they are compared both with the profiles in other parts of the city, and with the city as a whole.

In the case of the group of houses labelled P47.1–3, at Amarna, the artefactual evidence is totally dominated by tools and products relating to sculpture, leading to the unequivocal identification of this complex as a 'sculptor's workshop.' In this instance we can supplement the archaeological evidence with a scene in the tomb of Huya, high steward of Queen Tiye, at Amarna, in which a sculptor called Iuty is shown at work in his studio.

Many houses at Amarna, however, have produced artefacts suggesting that a wide range of activities was being conducted, perhaps by different members of

Systemic context	18th-Dynasty carpenter's tool-kit (see Aldred 1954, 688) **adze, axe, chisel, reamer, saw, bow-drill, mallet, awl, rubber** ↓ Removal of valuable bronze and large wooden items at abandonment of site. Potential loss of **adze**, **axe** and **saw** blades and handles. ↓
Archaeological context	Differential preservation of materials. Potential loss of wooden **mallet** and wooden handles of all other tools. ↓ Inadequate excavation. Potential loss of smaller items such as **chisel** and **drill**-head. ↓
Interpretive context	Inadequate recording, e.g. misidentifications or failure to record certain items. Potential loss of **rubber** and **adze**-handle (perhaps misidentified as grind-stone and stick respectively). ↓ Functional ambiguity, i.e. many carpenter's tools are easily confused with those of other crafts, e.g. **awl** (leather-working), **chisel**, **drill** and **mallet** (stone and metal working).

Figure 2.2 Flow-chart showing potential loss of objects as they pass from the systemic to interpretive contexts.

the household. The problem of interpreting such assemblages shows the potential gap between the world of textual labels and abstractions and the complexity of the real world as presented in the form of archaeological remains. A few other houses in the North and South Suburbs are also identifiable as workshops. House O49.14 is probably another sculptor's residence; Q46.23 was evidently occupied by a coppersmith; Q47.2 by a bead-maker; U35.2 by a painter and Q47.3 by a cobbler. The evidence, however, derives primarily from the interiors of houses (since these were the main target of pre-1970s excavators at the site), whereas the evidence of current excavations confirms that most craftwork actually took place in courtyards and open areas.

A group of houses (P49.3–6) were excavated by Ludwig Borchardt in his 1912 season at Amarna. The arrangement of these buildings, around a central courtyard, suggests that they may have constituted a workshop of some kind. When the P49.3–6 group was re-examined during the 1987 excavations at Amarna, it became clear that the courtyard, covered in a surface scatter of basalt chips, had not been excavated by Borchardt. The future excavation of such neglected open areas will no doubt provide a great deal more evidence concerning the plying of trades in individual households at Amarna.

TEXTUAL SOURCES

Another complicating factor, whether for visual or archaeological evidence, is our uncertainty as to whether 'private' patronage of professional craftsmen was a peripheral or dominant feature of the Egyptian economy. Similarly, it is difficult to tell whether the majority of skilled craftsmen plied their trades as private individuals in their own houses, or whether they were generally employed together in large state-controlled (or temple-controlled) workshops. When Eric Peet and Leonard Woolley were excavating parts of the city at Amarna in the early 1920s, they concluded that "within the main area of the city there is ... no evidence of the grouping of people of various classes or trades in different quarters of the town. High-Priest rubs shoulders with leatherworker, and Vizier with glass-maker" (Peet and Woolley 1923, 1–2). This rather intuitive quote not only suggests (incorrectly as it turns out, see Shaw 1996) that the grouping of houses of professionals and craftsmen at Amarna is largely random and unsystematic, but also indirectly raises the question of whether individuals in the city at Amarna, and other New Kingdom communities, can actually be identified socially by one specific skill or craft, or whether there were high degrees of multi-tasking by individuals (the ancient equivalents of 'portfolio' workers).

Fortunately, there are several New Kingdom textual sources, mainly deriving from western Thebes, which can help to illuminate the archaeological picture concerning trades and occupations. These sources may be divided into two basic types: (1) self-conscious literary and scribal documents, which can be used to explore the ancient Egyptians' own views on different trades and professions, and (2) administrative or legal records of specific individual accounts or administrative arrangements, which give some idea of the ways in which the population were divided up into different types of workers. It is the latter category that has the greatest potential from the point of view of identifying the degree to which particular craftworker's houses were grouped together. There are two administrative documents that are particularly relevant to this kind of analysis of craftsmen and professions: the *Wilbour Papyrus* (Gardiner and Faulkner 1941–52) and *Papyrus BM 10068* (Peet 1930b, 83–102).

The *Wilbour Papyrus* is an administrative document dating to Year 4 of Ramesses V (c. 1152 B.C.), which lists (for the purpose of assessing 'tax') a large number of plots of land in Middle Egypt. Each plot is described in terms of its size, calculated yield, the name of the priest or official who owns or administers it, the name of the nearest settlement and the name and occupation of the cultivator. David O'Connor used this source to plot the late New Kingdom patterns of settlement in Middle Egypt and occupation types. He distinguishes between zones with greater and lesser emphasis on agriculture and points out that different settlements have different proportions of the major types of 'occupation' listed: cultivator, priest, lady, herdsman, scribe, stable-master, soldier and 'Sherden' (East Mediterranean immigrant). In a pioneering study using the Statistical Package for Social Scientists on a body of textual data, Katary (1983) compared this raw data in the Wilbour Papyrus with the pro-scribal views expressed in such literary texts as *Papyrus Sallier II*, *Papyrus Anastasi VII* and *Papyrus Lansing* (see Gardiner 1937; Caminos 1954).

Papyrus BM 10068, one of a group of 12 'Tomb Robbery Papyri' recording judicial inquiries at Thebes and dating to Years 16 and 17 of Ramesses IX (c. 1115 B.C.), records on its *recto* a list of tomb robbers and stolen goods and on its *verso* a list of 182 households. The list of households is described as the "town register of the West of No from the temple of King Menmaare to the settlement of Maiunehes." Like the *Wilbour Papyrus*, *P. BM 10068* lists the names and occupations of a series of individuals. Peet determined that the listed houses lay in a line between the temples of Seti I, Ramesses II and Ramesses III and then turned west towards the contemporary village of Deir el-Medina (see Janssen 1992).

PROFESSIONS	%	PROFESSIONS	%
Priests	28.0	Herdsmen	10.0
Scribes	7.0	Fishermen	7.0
Medjay (police)	5.0	Coppersmiths	5.0
Administrators	4.0	Gardeners	4.0
Sandalmakers	4.0	Chief stablemen	3.0
Washermen	3.0	'Land-workers'	3.0
Beekeepers	2.0	Brewers	2.0
Attendants	2.0	Potters	2.0
Storemen	1.0	Porters	1.0
Incense roasters	1.0	Woodcutters	1.0
Doctors	0.6	Chief workmen	0.6
Guards	0.6	Goldworkers	0.6
Gilders	0.6	Measurers	0.6
Makers of 'w3t swi'	0.6		

Figure 2.3 Table listing the percentages of different occupations mentioned in Papyrus BM 10068.

Both *P. Wilbour* and *P. BM 10068* provide an indication of the relative importance of particular professions and trades. However, the high percentage of soldiers in *P. Wilbour* (43.7% according to Katary 1983) undoubtedly results from the nature of the document, since many military personnel would have been rewarded with plots of land. The high percentage of priests in *P. BM 10068* (28%) must also be somewhat unrealistic, since, as Peet (1930a, 84) cautions: "it is important not to lose sight of the very artificial composition of the population of the West of Thebes, where there was probably little business carried on except in connection with the long line of funerary temples of the kings and the Necropolis." On the other hand, the combination of both papyri provides an idea of the way in which an urban population might have been dominated by priests, soldiers, scribes and administrative officials. This equates with the evidence of the inscribed door-jambs of some of the largest houses at Amarna, since the titles of these wealthy householders imply that they would usually have been in charge of entourages of members of these four professions.

P. BM 10068 also includes many other occupations, of which the most numerous are gardeners, herdsmen, fishermen, coppersmiths and sandalmakers. Perhaps the most notable aspect of the lists is that they suggest that such trades were actually considered to be full-time professions. Beekeepers, brewers and gilders are relatively familiar categories of tradesmen, but the text certainly implies that 'incense roasters' and 'measurers' were also regarded as full-time workers in *P. BM 10068*. Figure 2.3, listing the different occupations (with percentages of the total population of the community), indicates that there were many types of occupation which might have left no trace in the archaeological record at Amarna, such as porter, guard, attendant or storeman.

It is possible that *P. BM 10068* lists the households in their actual topographical order. If this is the case, it would be possible to gain a text-based idea of the patterning of different trades within this particular community. If the document is a true reflection of the patterning within the community, despite the fact that it seems inherently unlikely for the settlement to have been strung out in a long row as the list-form would suggest, the different professions and ranks appear to be mixed together rather than separated into specialised zones. On the other hand, the distinctive 'neighbourhoods' identified archaeologically at Amarna (see Shaw 1996) may be echoed to some extent in occasional small groups of the same type of trade spread among the other types. It is to be expected, in an evidently temple-based community, that priests might group together (particularly in the immediate vicinity of the three royal mortuary temples), but there are also a few concentrations of other trades, in groups of up to five in a row (such as fishermen, herdsmen and sandalmakers). Most of the 'land workers' are concentrated at the southwestern tip of the settlement. There are also occasional pairs of the same profession, such as coppersmiths, scribes and brewers. Other occupations, however, are spread relatively evenly throughout the list.

A CASE-STUDY IN THE IDENTIFICATION OF CRAFTWORKERS' HOUSES: THE NORTH SUBURB AT AMARNA

If we return to the archaeological evidence at Amarna, it is possible to detect certain patterns in the city, in that there are some residential areas where the tools of particular trades are more common. The 'North Suburb,' which was excavated in four seasons between 1926 and 1932 (Frankfort 1927; 1929, Pendlebury 1931; 1932, Frankfort and Pendlebury 1933), consists of 298 buildings covering an area of about 300,000 square metres. This zone, located 600 metres to the north of the central city, is the only residential part of the city to have been almost entirely cleared by a single expedition (the Egypt Exploration Society). It should be noted however that about another 300 houses, in this northern zone, must have been lost either through modern cultivation and cemeteries or through the effect of the floodwater which created the wadis (one to the south of the suburb and the other cutting through the middle of it). The excavated area therefore represents only about half of the North Suburb's original population.

According to Frankfort and Pendlebury, the southern half of the North Suburb was subdivided into seven quarters: south-eastern, south central, eastern, northern, north-western, central western and southwestern. The 'northwest quarter' was identified as a possible merchants' quarter on the basis of the high numbers of corn-bins and magazines, leading to speculation that a quay may have been located nearby. The 'northern quarter' (the northeastern sector of the southern half, i.e. gridsquare U35) was tentatively suggested as a residential area of sculptors and artists working on the north tombs. The reasons cited for this theory are (a) the quantity of tools and debris; (b) the nearness of this quarter to the tombs themselves; and (c) the fact that these blocks of buildings appear to have been built by 'the same contractor.' A detailed examination of the published description (Frankfort and Pendlebury 1933, 31–7) shows at once that the incidence of bronze tools is very high. Nine out of the thirty-seven houses have yielded at least one adze, knife or axe, and in most cases there were several tools found together (such as the knife, adze and whetstone in U35.18 and the three bronze adzes in U35.16). In contrast, the 54 houses of gridsquare U36 have yielded only two axes altogether and no adzes at all; in the northwest quarter there are also no adzes. The northern quarter was evidently inhabited mainly by carpenters and joiners, as well as at least one painter (U35.2). However, this artefactual evidence does not by any means automatically add up to a community of official tomb-workers: the percentage of bronze tools among the artefacts of the Amarna workmen's village – usually considered to be a community of tomb-workers – is relatively low. The people in this quarter may equally well have been producing the many wooden fittings and items of furniture (such as columns, tables and beds) which originally graced most of the Amarna houses.

The very fact that these tools have been left behind in North Suburb houses (sometimes concealed beneath the floor) may indicate that these were tools owned by private craftsmen rather than the implements issued to government teams such as the Deir el-Medina workmen at western Thebes (since there is documentary evidence that a strict control was maintained over the issue and return of bronze tools, see Janssen 1975, 314). It is perhaps worth noting that three adzes were found under the floor of the house of Hatiay (T34.1 & 4) which was at the lower edge of the quarter north of the wadi. Hatiay is identified by his inscribed stone lintel as the 'Overseer of Works:' someone who might possibly have controlled or dealt with such a community of woodworkers as those in the northern quarter.

The western side of the suburb was identified, in a preliminary report, as a fishermen's quarter, on the basis of "the very large number of fish amulets and fish hooks" discovered there (Pendlebury 1932, 234). This theory seems quite plausible; an examination of the find-lists for the four western quarters confirms the relatively high occurrence of hooks and fish motifs compared with the eastern quarters (the houses in gridsquare U36, for instance, yielded no fish-hooks at all). However, there are sufficient other tools among the assemblages of houses on the western side of the North Suburb to indicate that fishing could have been simply a convenient sideline rather than necessarily a full-time occupation.

Many types of craftwork are present in the North Suburb as a whole, but the percentages of tools are low, and usually below average compared with the city as a whole. These figures might be interpreted as an indication that the inhabitants of the North Suburb were making a comparatively weak contribution to the economy of the city, but there is always the possibility that some of the inhabitants of the North Suburb plied some trade which is poorly reflected in the material remains. They may for instance (in view of their location immediately to the north of the city centre) have worked in some capacity in the large cluster of public buildings at the heart of the city. Some of the larger houses of the North Suburb may have housed scribes or clerks. There must, after all, have been a huge number of bureaucrats in the city at Amarna, since Akhetaten was, for a while at least, the official political and religious focus of an empire, stretching northeastwards into western Asia and southwards into Nubia. The 'Amarna letters' from the central city can only have represented a small fraction of the scribal activity generated by the processes of government, while the two major temples to the Aten in the central city must have been served by many priests and lesser officials answering to the chief priest Pawah (whom we know to have been housed in house O49.1 in the South Suburb).

Frankfort describes a block of houses (V36.7, .12 and .13) in the eastern quarter of the North Suburb as a tax-collector's official residence, basing his interpretation on the rather slim evidence of a set of magazines fronted by a portico with a dais, which he identifies as offices (Frankfort and Pendlebury 1933, 30–31). However, just as the many silos and magazines of the north-western quarter strongly support some form of trading or redistributive role for these houses, so the unusual outbuildings of the houses in the eastern and south central quarters of the North Suburb may constitute architectural indications of professions which the finds inside the houses can only hint at.

The percentage of textile remains recorded from the North Suburb, although low in absolute terms, is higher than the average for the whole city, yet the proportion of spinning and weaving artefacts is well below that found either in the South Suburb or in the North City. This dislocation between centres of production and consumption again suggests a city in which products were often traded not only between adjacent neighbourhoods but also from one end of the city to the other. This combination of low productivity and high consumption would also tie in with the possible high percentage of bureaucrats living in the North Suburb.

Further enlightenment can probably also be found in an unusual architectural feature of the North Suburb. The profile of house-sizes in the North Suburb is similar to that of the South Suburb with regard to smaller and medium-sized houses, but the missing element in the North Suburb was the élite: the wealthiest members of Amarna society (houses above 400 m^2 in area). It seems almost inevitable that craftwork would have thrived in the South Suburb where the households of very high officials such as the vizier Nakht, the high-priest Pawah and the general Ramose were located. These men and their households would doubtless have required all kinds of artistic and ornamental products. In the North Suburb, however, the more subdued level of craftwork probably reflects the comparative dearth of such élite officials (most of whom had naturally taken up residence in the earlier southern suburb). If the inhabitants of the North Suburb were not catering for the needs of the élite then they were perhaps middle-ranking scribal officials or 'managers,' at least partly employed by state or temple.

CONCLUSIONS

A number of conclusions can be drawn from the above discussion of visual, textual and archaeological evidence for the practising of different trades and crafts. First, it is often very difficult to reconcile textual and archaeological data concerning crafts and professions. Secondly, many publications discussing New Kingdom villages, towns and cities present only part of the picture, since streets, courtyards and 'empty' public spaces have tended not to be excavated in the past. Thirdly, the evidence from individual households at Amarna suggests that most households contained various individuals plying a diverse range of crafts rather than simply indicating the principal 'job' of the head of the household.

This situation is often even more complicated when we consider the very large villas at Amarna, where both family members and servants would be plying different trades, and in addition surrounding small households (perhaps linked to the large ones by the need to obtain water and grain) probably have to be treated as extensions to the major 'nuclear' house. In other words, our pursuit of the houses of individual craftsmen and professionals is complicated by the fact that no working person in an urban context can be examined independently of the community within which he or she is working. Each urban household must be interpreted in terms of networks of production and consumption both on neighbourhood and city-wide levels.

This paper began with questions relating to the archaeological recognition or 'definition' of an oil-boiler. It ends with no easy solution to that particular problem, but the case-study of the Amarna North Suburb suggests that it will sometimes be possible to hypothesise the work practices of certain groups of individuals by examining the total excavated information, particularly in the context of neighbouring houses and suburbs. As we accumulate information on a greater variety of types of dwelling in larger numbers of New Kingdom towns, it might well become possible to identify not only such esoteric inhabitants as oil-boilers and incense-roasters, but also that quintessentially Egyptian professional: the bureaucrat.

REFERENCES

Aldred, C., 1954, A History of Technology, Oxford.
Caminos, R. A., 1954, *Late-Egyptian Miscellanies*. Brown Egyptological Studies, London.
Davis, W., 1983, Artists and patrons in predynastic and early dynastic Egypt, *Studien der Altägyptischen Kultur* 10, 119–40.
Eyre, C., 1995, Symbol and Reality in Everyday Life, in C. Eyre (ed.), *Abstracts of the 7th International Congress of Egyptologists, Cambridge, 3–9 September 1995*, 55–56. Oxbow Books, Oxford.
Frankfort, H., 1927, Preliminary report on the excavations at Tell el-'Amarnah, 1926–1927. *Journal of Egyptian Archaeology*, 13, 209–18.
Frankfort, H., 1929, Preliminary report on the excavations at el-'Amarnah, 1928–29, *Journal of Egyptian Archaeology*, 15, 143–9.
Frankfort, H. and Pendlebury, J.D.S., 1933, *City of Akhenaten II*. Egypt Exploration Society, London.
Gardiner, A. H., 1937, *Late-Egyptian Miscellanies*. Bibliotheca Aegyptiaca 7, Brussels.
Gardiner, A. H. and Faulkner, R. O., 1941–52, *The Wilbour Papyrus*. 4 vols. O.U.P., Oxford.
Janssen, J. J., 1975, *Commodity prices from the Ramessid period: an economic study of the village of necropolis workmen at Thebes*. Brill, Leiden.
Janssen, J.J.,1992, A New Kingdom settlement: the verso of P.BM10068, *Altorientalische Forschungen* 19, 8–23.
Katary, S. L. D., 1983, Cultivator, scribe, stablemaster, soldier: the Late Egyptian Miscellanies in the light of Papyrus Wilbour. *The Ancient World*, 6, 71–94.
Peet, T.E., 1930a, Two letters from Akhet-aten, *Liverpool Annals of Archaeology and Anthropology*, 17, 82–97.
Peet, T.E., 1930b, *The great tomb-robberies of the 20th Egyptian dynasty*. O.U.P., Oxford.
Peet, T. E. and Woolley, C. L., 1923, *City of Akhenaten I*. Egypt Exploration Society, London.
Pendlebury, J. D. S., 1931, Preliminary report of excavations at Tell el-'Amarnah 1930–31, *Journal of Egyptian Archaeology*, 17, 240–43.
Pendlebury, J.D.S., 1932, Preliminary report of excavations at Tell el-'Amarnah 1931–32, *Journal of Egyptian Archaeology*, 18, 143–149.
Shaw, I., 1996, Akhetaton – the city of one god, in J. Goodnick Westenholz (ed.), *Royal cities of the Biblical World*, 83–112. Bible Lands Museum, Jerusalem.
Wente, E., 1990, *Letters from ancient Egypt*. Scholars Press, Atlanta.

Chapter 3

Canaan in Egypt: archaeological evidence for a social phenomenon

Rachael Thyrza Sparks

Abstract
In the 2nd millennium B.C., Canaanites were introduced into Egyptian society through a variety of means: brought back from military campaigns as living booty, or drawn to the area for employment as mercenaries, traders or workers with a variety of specialist skills. We can trace these movements through the textual record, which shows people bearing Semitic names engaged in different occupations throughout Egypt, and other related phenomena such as the introduction of foreign loanwords and Asiatic cults. But how far can this social phenomenon be traced through the archaeological record? This paper attempts to draw together the information currently available, and consider some of the broader methodological issues, such as the kinds of material culture markers that might alert us to foreigners living amidst Egyptians, and how to identify the social and technological impact of their presence on Egyptian culture.

INTRODUCTION

Interaction between Egypt and the Levant took a variety of forms during the course of the 2nd millennium B.C. Egyptian foreign policy with regard to her northern neighbours may have begun with a period of targeted commercial and occasionally military interest (Cohen 2002, 50). Meanwhile, on the other side of the border, a growing Asiatic presence in the Eastern Delta provided the framework for the ultimate rise to power of the 15th Dynasty rulers (Bietak 1996). A period of more intensive economic and cultural involvement between these regions followed. While subsequent events lead to the defeat of the Hyksos, this involvement continued during the Late Bronze Age, directed by a more aggressive Egyptian foreign policy that was reflected in changes in Canaanite material culture at both the stylistic and technological level (Liebowitz 1986; McGovern 1989; Mumford 1998; Higginbotham 2000). At the same time there was a corresponding flow of Canaanite goods, captives and professionals back to Egypt that left its own mark on Egyptian society and culture. This paper will examine the nature of this exchange, how it can be detected in the archaeological record, and the impact it had on the development of Egyptian crafts and products.

The concept that foreigners may have been resident in Egypt during the Middle Kingdom and later periods is not new; Petrie first proposed this idea back in the 1890s to explain some of the material he was uncovering at sites such as Lahun and Gurob (Petrie 1890; 1891). However attention was focused on possible links with Aegean cultures, and a Levantine connection was barely considered. Thus when Petrie discovered Palestinian-style toggle pins at Gurob, they are described as "curious pieces...like those found in one class of Cypriot tombs" (Petrie 1891, 19). This is hardly surprising: many Egyptian sites were excavated when the archaeology of the Levant was in its infancy and Canaanite material culture poorly known and understood. This situation has changed markedly since the late 19th century, with extensive publication of hundreds of sites across Syro-Palestine, and detailed studies of object classes reflecting Canaanite lifestyle and technology providing a background against which to re-examine material of foreign origin found in Egypt itself. Added to this are more recent excavations across northern Sinai and at sites such as Tell el-Dab'a, Tell el-Maskhuta and Tell Heboua in the Eastern Delta (Oren 1987; Bietak 1996; Holladay 1982; 1997; Redmount 1995; El-Maksoud 1998). These sites often served as interfaces between Egyptian and Canaanite cultural spheres, and have provided a wealth of new material that can be used to reassess the issue.

TEXTUAL EVIDENCE FOR CANAANITES IN EGYPT

As contacts with the north intensified, depictions of Canaanites in Egyptian art increased. Commercial involvement, including trade by land and sea, is suggested by Middle Kingdom scenes at Beni Hasan and Sinai, and in the 18th Dynasty tomb of Kenamun at Thebes (Newberry 1893, Pl. XXXI; Bietak 1996, Fig. 13A–B; Davies and Faulkner 1947). Asiatic soldiers appear as individuals and groups in the Egyptian army, pointing to the presence of foreign mercenaries from the Middle Kingdom onwards (*e.g.*, Newberry 1893, Pls. XVI, XLVII; Redford 1988, 15–17; Bietak 1996, 19, Fig. 14). Finally as the Southern Levant was absorbed into the Egyptian empire from the mid-18th Dynasty, we see a new range of depictions emerge: conquest scenes depicting bound Asiatic prisoners being led before the pharaoh, Canaanite 'tribute' being brought to Egypt, and foreign princes paying homage (Redford 1988, 13–15, 17–18; Aldred 1970; Warburton 1997, 141; Pritchard 1969, 248–9). From this time, Canaanite figures are also used as generic symbols of Egyptian conquest and power, as illustrated by the many cosmetic and personal items in the tomb of Tutankhamen with bound Asiatic captives used as decorative motifs (James 2000, 195, 270–3).

Another source of information about the ethnic makeup of communities in Egypt comes from a study of administrative documents listing people's names

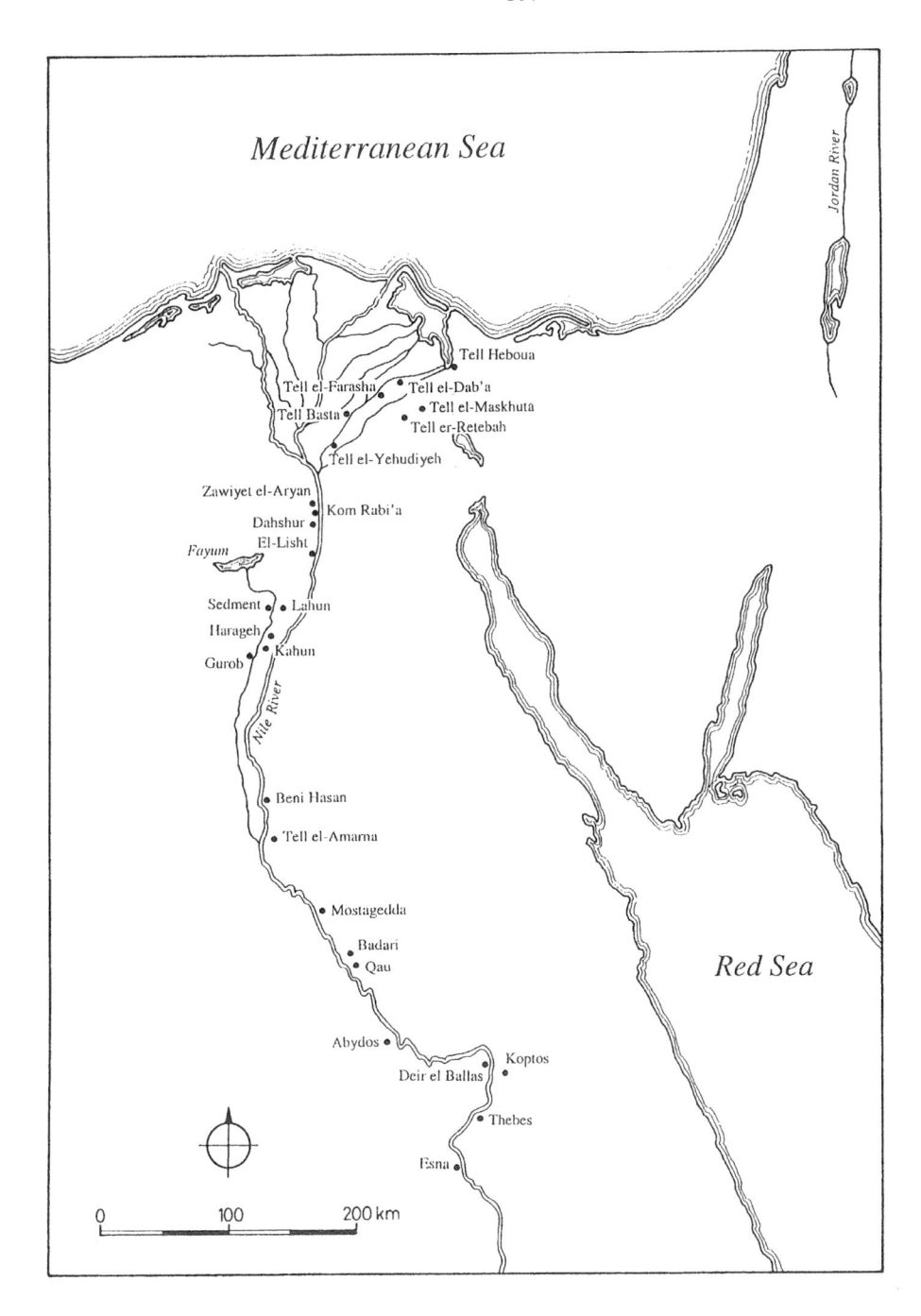

Figure 3.1 Map of Egypt showing sites discussed in the text.

and occupations. Quite a large amount of data has now been assembled on this topic (Schneider 1992), although identifying names of foreign origin in Egyptian texts is not without its problems (Ward 1994, 63–65). Through such documents it is possible to trace the appearance and role of foreigners in Egyptian society. They show people bearing Western Asiatic names beginning to appear in the textual record sometime during the 12th Dynasty. Canaanites may also be identified in Egyptian texts through the use of epithets such as 'the Asiatic' ('3m, '3mt), or a determinative sign marking a foreign name (Ward 1994, 63). On some occasions, individuals labelled in this way actually carry Egyptian names, pointing to the beginnings of integration into Egyptian society (David 1986, 190). The soldier Ahmose, whose autobiography is so important to understanding the defeat and expulsion of the Hyksos, has an Egyptian name, but the names of his parents are Asiatic (Leahy 2000, 231). On other occasions, a person might be known by both an Egyptian *and* a foreign name – as was the case with Paheqamen/Benya, chief architect and chief of the treasury in the reigns of Hatshepsut and Tuthmosis III. His parents carry Asiatic names, while he is portrayed in his tomb as fully Egyptian in appearance (Ward 1994, 64; Leahy 2000, 233). An Egyptian name may have been assumed deliberately as a means of blending into Egyptian society, and second or third generation Canaanites may have considered themselves to be Egyptian, even when the authorities did not. These individuals seem to have had varied roles, with occupations ranging from servants, cooks, gardeners, workmen, and weavers to high ranking positions in the Egyptian administration (David 1986, 190; Ward 1994, 67; Leahy 2000, 229).

ARCHAEOLOGICAL EVIDENCE FOR CANAANITES IN EGYPT

It is clear from textual sources that Canaanites were present in Egyptian society from the Middle Kingdom, and that they continued to be a significant element down to the end of the Ramesside period. It remains to be determined whether this presence can be successfully isolated in the archaeological record, and whether resident Canaanites can be distinguished from Egyptians and other groups through a study of their material culture remains. This is by no means a simple process. It must first be established whether there are any cultural markers that can define Canaanite 'ethnicity' in Egypt, through a comparison of the artefact assemblages found in Egypt and the Levant. Then it must be determined whether the foreign-style objects that have been discovered in Egypt were in fact being used by Canaanites, and can be said to therefore be a part of their 'foreign' identity, rather than goods traded into the region for use by the wider population.

The best method of determining this would be to look at a range of evidence, and to consider which aspects are likely to reflect a distinct, non-Egyptian lifestyle. This might include consideration of factors such as whether buildings or tombs show any distinctively Canaanite design elements or construction methods (*e.g.*, Bietak 1992); how domestic and work space is defined in the household, and what dietary habits are practised there. This sort of analysis has rarely been attempted for Egyptian settlement sites, requiring some form of spatial and contextual analysis of finds, along with detailed archaeobotanical and faunal studies of household rubbish. However the more careful excavation techniques that have been practised more recently at sites such as Memphis, Tell el-Dab'a, Tell el-Maskhuta, and Tell el-Amarna may well allow this sort of analysis to be carried out in the future. Other areas where it might be possible to detect foreign settlers might be in a study of artefact assemblages: the tool sets and methods employed by various specialist crafts, such as weavers and woodworkers; objects reflecting daily life, the type of religious activities that are conducted, in both public and private spaces, and finally, the way in which the living deal with death.

When evaluating any object for its possible use as an ethnic marker, several points should be considered. Does the object have a function that is closely tied to a particular cultural milieu, or would it be equally at home in another culture? Is it made of a material that would make the object desirable irrespective of its function? Objects that carry luxury or prestige value are particularly adept at crossing cultural boundaries, and therefore may be less useful indicators of the presence of foreigners. Finally, the archaeological context in which the object was found may indicate whether the object was used in the same way in both Egypt and the Levant, another clue to the possible ethnic origin of the owner. Associated material is particularly important here, as an object that is found with other Canaanite goods is less likely to be a random, or accidental occurrence, and more likely to reflect deliberate choice.

How many of these elements are visible in the archaeological record depends upon whether a Canaanite immigrant was isolated in the community or part of a larger Canaanite network there. This may have considerable influence on their ability to continue to live a 'Canaanite' lifestyle. If isolated, they are unlikely to be able to call on people trained in Canaanite technology to provide them with their daily needs in terms of housing, clothing, food or goods, and are more likely to obtain these things from local Egyptian markets. If there is an existing group of Canaanites living in the region, it is more probable that some elements of their foreign lifestyle could be preserved if desired. Large groups of people may therefore be easier to detect than individuals.

PERSONAL APPEARANCE

Egyptian art used standard conventions for depicting foreigners, and thus the characteristics of the 'Nubian,' 'Libyan,' 'Hittite' or 'Asiatic' can be reduced to a number of significant features, including hairstyle, facial type, skin colour, tattoos, clothing, jewellery or weapons (Figures 3.2, 3.4a, 3.6d; Leahy 2000, 226). These show that Egyptians considered northerners to have a distinctive physical appearance, characterised by particular styles of grooming, clothing and personal adornment. While these representations are highly standardised and idealised, many of the features shown are confirmed in part by depictions of Canaanites

Figure 3.2 Canaanite women, Beni Hasan Tomb 3 (after Newby 1980, p. 38).

from Near Eastern sources such as cylinder seals, inlays and figurines. Some of these are also suggested by sets of artefacts that may be considered indicative of Canaanite habits of dress, and which can be isolated in the archaeological record.

Clothing

Egyptians tended to wear linen garments tied in place or secured with sashes (Vogelsang-Eastwood 2000, 286–90). Although they made occasional use of coloured bands or embroidered details, these were predominantly plain: less than 3% of the textiles recovered from Tell el-Amarna show any trace of colour (Kemp and Vogelsang-Eastwood 2001, 152). Canaanites, on the other hand, are often depicted in Egyptian art wearing highly patterned garments (Figures 3.2, 3.4a; Brovarski *et al.* 1982, 180; *cf.* Negbi 1976, Fig. 103). Even when plainer versions appear, they frequently have decorative coloured borders (Redford 1988, 21). Many of these garments may have been made from wool, as contemporary texts indicate that dyed wool was a popular tribute and trade item in the Near East (Cochavi-Rainey 1999, 181–3, RS 16.146 + 161; 187–9, RS 17.227 and KTU 3.1; 191–3, RS 11.732; Lilyquist 1998, 215 and references *infra*). This popularity did not seem to extend to Egypt, where wool is rare in both burial and settlement contexts.

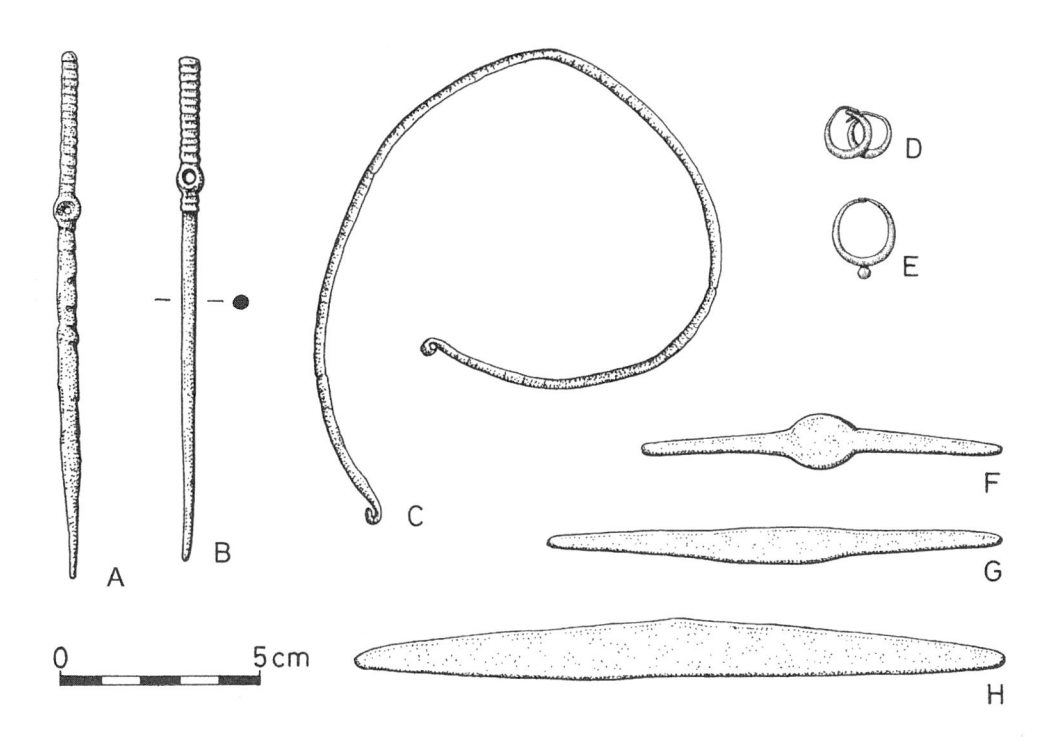

Figure 3.3 Canaanite style clothing accessories and jewellery found in Egypt. a) Metal toggle pin, Tell el-Maskhuta (after Holladay 1997, pl. 7.9.19); b) copper toggle pin, Tell el-Dab'a (after Bietak 1991, fig. 69.3); c) silver torque, Tell el-Maskhuta (after Holladay 1997, pl. 7.9.6); d) gold earrings, Badari (after Brunton 1930, pl. XXXV.34); e) gold earring, Sedment (after Petrie and Brunton 1924, pl. XLII.8); f–h) gold headbands, Tell el-Maskhuta (after Holladay 1997, pl. 7.22D).

One distinctive element associated with Canaanite dress is a type of fastener known as the toggle pin. These are characterised by an eye one third to midway down the shaft, sometimes used to attach a cylinder seal or scarab, and a pointed end that could be used to fix garments in place (Ziffer 1990, 59*–60*; Kenyon 1965, 571). They were most commonly made of bronze, but more expensive silver, gold and electrum versions are known, as well as the occasional bone or ivory example (Henschel-Simon 1937). Toggle pins were widespread across Canaanite society during the Middle and Late Bronze Age, appearing in élite and middle ranking contexts that include domestic as well as funerary settings. Discoveries of these pins on undisturbed bodies in the burials at Middle Bronze Age Jericho suggest that they were used to fasten garments in a variety of ways – appearing on the shoulder, chest, and waist (Kenyon 1965, 473, 567). A study of their frequency in tombs suggests that they were not an invariable part of Canaanite dress; at Jericho, for example, only one in four bodies seems to have been equipped with a toggle pin (Kenyon 1965, 567). However at Pella, further to the north, this ratio seems to be much higher. Whether there is a gender bias operating, or whether it was common for some groups to use several pins, perhaps reflecting warmer or more elaborate clothing, is difficult to establish in a period and region where multiple successive interments were common and many burials consequently disturbed.

In Egypt, the situation is somewhat different, with finds of toggle pins being comparatively rare. Examples are most common in tombs at northern sites such as Tell el-Dab'a, Tell el-Maskhuta, Tell el-Yehudiyeh, Tell Heboua, Kom Rabi'a and Gurob (Figures 3.3a–b; Bietak 1991, Figs. 69.3, 154.1–3, 120.2a; Holladay 1997, Fig. 7.9.19–20; Petrie 1906, Pl. VI.10–14; El-Maksoud 1998, Fig. 46; Giddy 1999, 168, Pls. 36.88a–c, 2023; Petrie 1891, Pl. XXII.1–3). While evidence for their placement is limited, there are signs that the pattern matches that found in Palestine; on a body from Tell el-Yehudiyeh, the pin was positioned on the left shoulder, and this positioning was also noted at Tell el-Dab'a (Tufnell 1978, 87, Fig. 9.11–16; van den Brink 1982, 45; Philip 1995, 73). Toggle pins also appear in an Egyptian domestic setting (*e.g.,* Giddy 1999, 168). While some local manufacture seems probable, toggle pins are also known to have been imported from the Near East, as when King Tushratta sent gold and silver examples as part of his daughter's dowry (Moran 1992, EA 25: I.22–32, III.56, 64). In all these instances the style and size of the pin, as well as the way in which it was used, seems to mirror contemporary Canaanite practice. These facts, plus the general rarity of the type, suggest that where toggle pins do occur they represent an introduced style of dress that was never widely adopted in Egypt, and which therefore may be indicative of resident northerners.

Adornment

Personal adornment is another area where distinctions may be drawn between Canaanite and Egyptian customs. The metal 'torque' was a simple form of Middle Bronze Age necklace that has been found at a number of north Levantine sites, including Ras Shamra and Byblos (Frankfort 1927, 150; Schaeffer 1949, 48–128; Tufnell and Ward 1966, 208–211). A handful of examples have also been discovered in Egyptian contexts, including graves in the Fayum, Mostagedda, Deir el-Ballas, Abydos, Tell el-Maskhuta, and in a domestic setting at Lahun (Figure 3.3c; Frankfort 1927, 149; Petrie 1891, Pl. XIII.18; David 1996, 135; Holladay 1982, 45; 1997, Fig. 7.9.6). It should be noted that although these are sometimes assumed to belong to women (David 1986, 135), torques appear to have been worn by both sexes in the Levant, suggesting that this type of jewellery was not gender specific (*e.g.,* Negbi 1976, *cf.* Cats. 72 and 1546).

Another introduced custom may be the use of decorative sheet metal for personal adornment. Narrow strips of gold, silver and electrum are known from tombs and scrap metal hoards in Middle and Late Bronze Age Palestine, where they have usually been identified as headbands (Maxwell-Hyslop 1971, 120–122; Ziffer 1990, 57*). Comparable material appears in contemporary tombs at sites such as Tell el-Dab'a, and Tell Maskhuta (Figures 3.3f–h; Bietak 1996, Pl. IIA; van den Brink 1982, 35; Holladay 1997, Pl. 7.22D). Other popular sheet metal appliqués include rosettes and discs, pierced for sewing onto cloth (*e.g.*, Petrie *et al.* 1952, Pl. A.5–6; James and McGovern 1993, 150–1). Similar appliqués were attached to cloth found in the tomb of Tutankhamen (Cochavi-Rainey 1999, Fig. 27), showing the continuation of this style of adornment down into the later 18th Dynasty. Another Asiatic custom that was popularised during the 18th Dynasty was the wearing of earrings (Brovarski *et al.* 1982, 227). While Egyptians soon develop their own distinctive versions, including various form of ear studs, simple drop-lunate earrings of the type popular in the Levant appear at sites such as Abydos, El-Ahaiwah, Badari and Tell el-Dab'a and may reflect more Canaanite tastes (Figures 3.3d–e; Randall-MacIver and Mace 1902, Pl. LIII; Brovarski *et al.* 1982 233, Cat. 305; Brunton 1930, Pl. XXXV.34; Bietak 1991, Fig. 17.2).

Personal grooming
Combs are rare in Levantine assemblages, probably because the majority were made of perishable materials such as wood. This impression is confirmed by finds from the Jericho tombs, where unusually favourable preservation of organic remains led to the survival of a number of wooden combs dating from the MB IIB–LB I period (Figure 3.5a; Kenyon 1965, Figs. 142, 188, 243). These appear in around 40% of burials of this period, and follow a standard design with the back carved to form three to four peaks along the length of the comb with the body cut away below to form a row of decorative 'windows.' These combs are quite small, usually 40–65 mm in length. Many were found in association with skulls or wigs, while some were placed in baskets containing other cosmetic items (Kenyon 1965, 473, 499, 574). Very similar wooden and ivory combs have also been found in Egyptian tombs at sites such as Tell el-Dab'a and Lahun (Figures 3.5b–c; Bietak 1991, Fig. 19.1–2; Petrie 1890, Pl. VIII.31; Petrie Museum UC 7097 and 7099). This is exactly the kind of personal item that could have been brought to Egypt with migrants, and it would be interesting to see whether these were made from local or northern wood types.

LIFESTYLE

Business activities
Canaanite merchants and business men were clearly active in Egypt. One way of detecting such activity could be through the presence of objects reflecting foreign weights and standards, as these imply the need to calculate by alternative systems of measurement and hence point to the involvement of individuals outside the Egyptian administration (Heltzer 1994). A scene from the 18th Dynasty tomb of Kenamun shows one setting where this sort of interaction took place, with Syrian sailors offloading a cargo of amphorae onto the quayside, where Egyptian merchants are seated with balance scales, ready to carry out some business (Davies and Faulkner 1947, Pl. VIII).

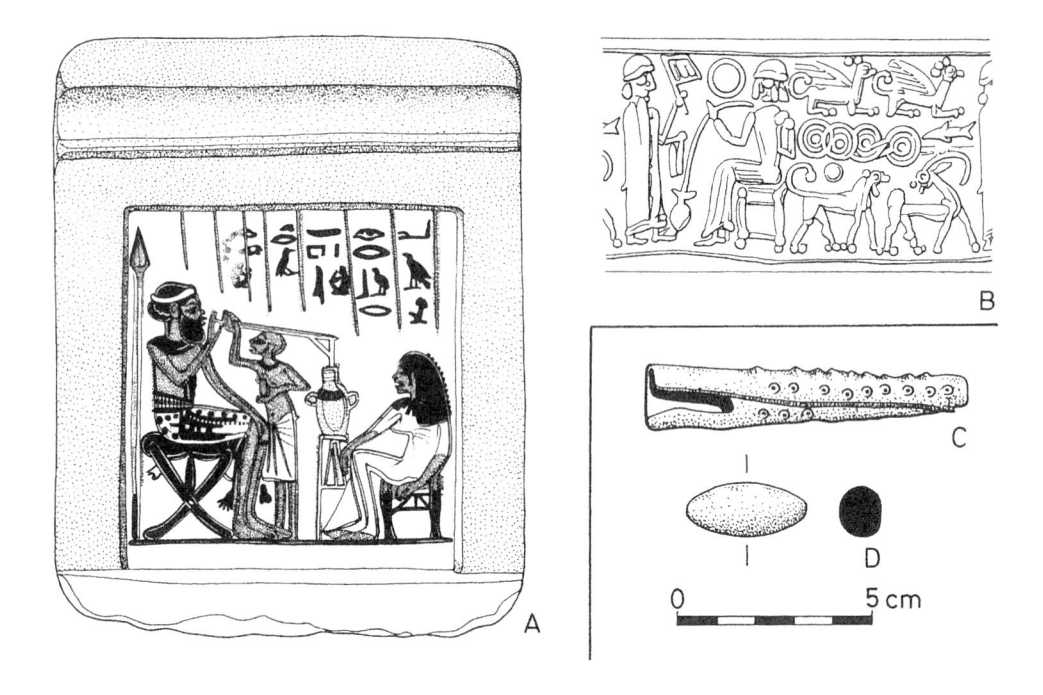

Figure 3.4 Objects associated with drinking and business. a) Syrian soldier stela (after Arnst et al. 1991, p. 129); b) cylinder seal, Ugarit (after Amiet 1992, fig. 12.54); c) strainer tube, Tell er-Retabeh (after Petrie 1906, pl. XXXVB left); d) spherenoid weight, Kom Rab'ia (after Giddy 1999, pl. 42.1806).

Petrie claimed that foreign weights made up around 83% of the total assemblage at 18th Dynasty Lahun, and at least 21% at New Kingdom Gurob (David 1986, 173; Petrie 1926, 6). While Petrie may have been over enthusiastic in attributing the function of a balance weight to some of his objects, there are a number of clearly defined weight types which seem to support his argument. One such is the spherenoid or lozenge-shaped weight, typified by a convex back, flat cut ends and narrow flattened base (Petrie 1926, 6, 'barrel'). These are common throughout the Late Bronze Age Levant, and may represent a local system of measurement, appearing in a range of graded sizes in materials such as hematite, basalt and limestone, sometimes identified by simple marks on the base (*e.g.*, Petrie *et al.* 1952, Pl. XXII.62.12–13, 37, 39, 42, 47–8, 62–3; James and McGovern 1993, Figs. 127.4–6, 128.2; Schaeffer 1962, Fig. 63E; Pulak 1997, 247–8, Fig. 18). In Egypt, several examples of this type of weight have been found at sites such as Kom Rabi'a, Gurob and Tell el-Amarna (Figure 3.4d; Giddy 1999, Pl. 42.1806, 220, 2055; Thomas 1981, I.71 No. 542; II, Pl. 53.542; Frankfort and Pendlebury 1933, Pl. XXVIII.1).

Food and drink

The preparation and consumption of food and drink is another area where it may be possible to detect a characteristically Canaanite lifestyle. This may be reflected by specific types of object assemblages, and food remains that survive in kitchen dumps or as part of ritual or funerary offerings. Some cooking and medicinal plants appear to have been introduced into Egypt from the Near East during the 18th Dynasty or earlier. They may have been brought back to Egypt by Egyptians who had spent time in the Levant, by Levantine immigrants, or through trade networks that opened up as a growing Canaanite community created the demand for foreign herbs and fruits. New types include black cumin and safflower, which was used both for making cooking oil and as a dye (Germer 1998, 88–9). Another 18th Dynasty introduction was the pomegranate (Brovarski *et al.* 1982, 113–4, 167). In Egypt, this seems to have been adopted as a food, in offerings and as a vermifuge agent, while the shape quickly gained popularity in the decorative arts (Brovarski *et al.* 1982, 114, Cats. 99, 187-188; Aston 1994, Type 148; Vandier d'Abbadie 1972, Cat. 809; Ayrton *et al.* 1904, Pls. XV.16, XVI.6 right, LV.19; Petrie and Brunton 1924, Pl. LXI.66; Quibell 1901, 142). However while the pomegranate had a strong association with fertility in its Levantine setting, it is not clear whether this symbolic significance was transferred to Egypt along with its visual form.

The introduction of these types of products may indicate the presence of Canaanites with distinctive dietary habits, or simply that Egyptians were becoming more cosmopolitan in their own tastes. A more useful indicator of cultural identity may lie in artefact types that are linked to a culture-specific activity. One such might be objects relating to the consumption or wine or beer. An unusual stela from Tell el-Amarna provides some clues concerning Near Eastern practices in this respect (Figure 3.4a; Brovarski *et al.* 1982, 109, Fig. 34; Arnst *et al.* 1991, Cat. 80, Berlin 14122). This depicts a man and his wife seated before an amphora, with a servant standing between them holding a juglet and directing the mouth of a long drinking straw towards the man. The couple are given the Semitic names of *Trr* and *Irbr* in the accompanying inscription (Ward 1994, 62, nn. 5, 71). However while the man is given the characteristics of an Asiatic – beard, hair fillet and highly coloured, patterned clothing – his wife and servant are shown in Egyptian dress. A spear is propped up against the wall behind the man, leading to the suggestion that he may have been a soldier.

The drinking straw first appeared in Mesopotamia, where it was used to help filter beer. By the Late Bronze Age it had also spread to the Levant, where scenes on cylinder seals show it in use by both men and women (Figure 3.4b; Petrie 1931, 7, Pl. XIII.33; Collon 1982, 38–9, Cat. 7; Amiet 1992, Figs. 4.8, 12.54, 45.250; Oates *et al.* 1997, Fig. 72). A number of perforated metal tubes found there have been identified as the strainer tips for these straws (*e.g.*, Woolley 1955, 281, Pl. LXXIII.AT/8/26; Oates *et al.* 1997, 116, Fig. 235.61, 63; Petrie 1934, Pl. XXXII.423). The practice was subsequently introduced into Egypt, where a handful of metal strainer tips have been found dating from the 18th Dynasty onwards at sites such as Tell el-Amarna, Tell er-Retabeh, Gurob and Koptos (Figure 3.4c; Griffith 1926, Fig. 1; Petrie 1906, Pls. XXXIVA.24, XXXVB; 1914, 38, Pl. XLIV.134–5; Petrie Museum UC 7776; for additional unprovenanced examples, see Petrie Museum UC 34514–5). These were sometimes misidentified as graters or rasps (Griffith 1890, 46; Petrie 1906, 32–3; 1914, 38). Twentieth Dynasty examples from anthropoid sarcophagi burials at Tell el-Yehudiyeh were found sitting inside ceramic amphorae, similar in shape to that depicted on the Berlin relief (Griffith 1890, 46, Pl. XV.20–21; for the date of the ceramic type, see Aston 1996, 66, Fig. 204a). One strainer even has remnants of the original reed straw preserved inside (Petrie Museum UC 64827). As Egyptians are almost never depicted using drinking straws themselves, and as finds of the associated fittings remain in the minority, it seems probable that this practice was never widely adopted in Egypt, and thus may have continued to be indicative of Astiatics.

Other household goods

One of the more common finds in settlement and burial contexts in MB II–LB I Palestine are bone inlays featuring a range of geometric designs incised and inlaid with black pigment to stand out from the pale colour of the bone (Liebowitz 1977). Common motifs include the ring and dot design, various forms of hatching, zigzags and the guilloche. While most inlays are rectangular, there is also a group of silhouette forms, cut out in the shape of various types of animal or geometric forms. These inlays are thought to have been used to decorate the sides and lids of wooden boxes, although the wood itself rarely survives. The strips were pierced for attachment by pegs, while the silhouettes were glued in place. These inlays are characteristic of South Levantine assemblages but much more rare in neighbouring areas, where they appear in a more restricted range of designs (Liebowitz 1977, 91–2).

Although boxes with bone or ivory inlays are not unknown in Egypt, material of comparable style has a limited distribution and would appear to represent some sort of Canaanite influence, lasting from the late Middle Kingdom through to the early 18th Dynasty. This influence may have travelled in both directions for a time, as some Egyptian motifs appear back in the Palestinian repertoire, with rare pieces such as the El-Jisr inlays having a more marked 'egyptianising' style (Amiran 1977). In general however, the 'Canaanite' inlays found in Egypt tend to be very simple in design. Examples are known from sites such as Tell el-Maskhuta, Tell el-Yehudiyeh, Zawiyet el-Aryan, Abydos, Esna, Gurob, Sawama and Sedment (Figures 3.5d–f; Holladay 1997, Pls. 7.22E; XIID.411; Dunham 1978, 37; Ayrton *et al.* 1904, Pl. LVIII.22; Downes 1974, Fig. 99.307E and 163E; Brunton and Engelbach 1927, Pl. XXI.42; Bourriau and Millard 1971, Pl. XVIII.3; Petrie and Brunton 1924, Pls. VI.20, LXIII.1723a; Petrie Museum UC 18858).

An analysis of two Egyptian cemeteries gives an idea of the frequency of this kind of material. Inlay strips were found in 11 of the 299 published tombs from Esna, ranging in date from the Middle Kingdom through to the early 18th Dynasty, a total of around 4% (Downes 1974). Where decoration is mentioned, it

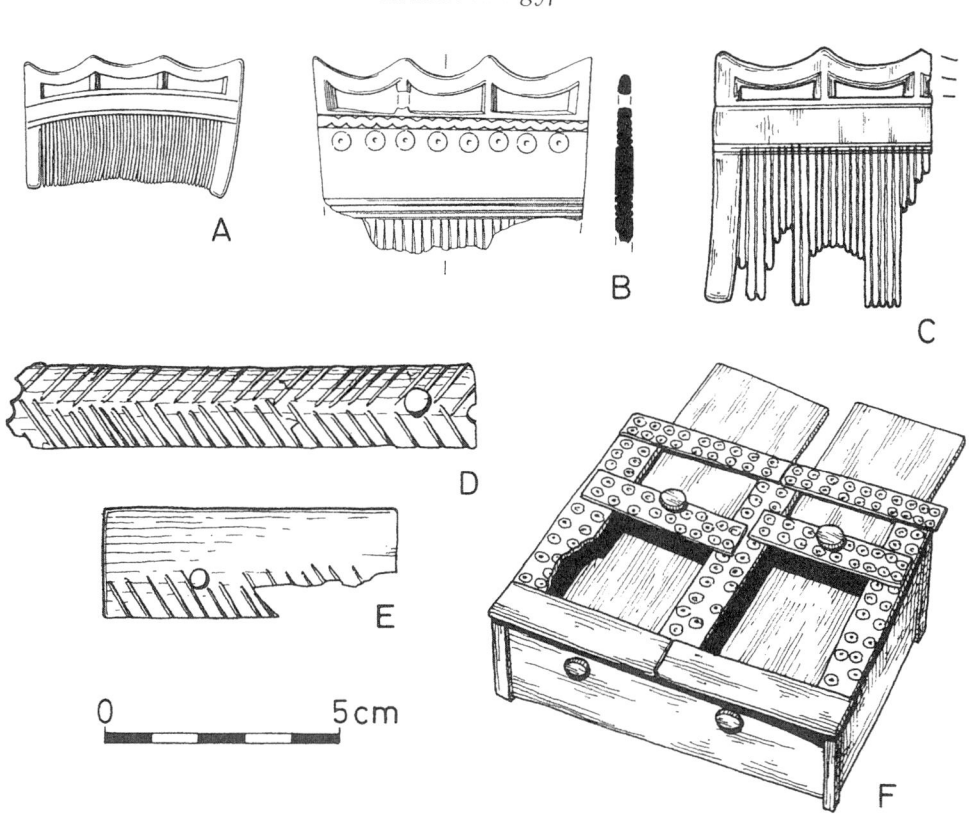

Figure 3.5 Canaanite combs and box inlays. a) Jericho (after Kenyon 1965, fig. 243.5); b) Tell el-Dab'a (after Bietak 1991, fig. 19.1); c) Lahun (after Petrie 1890 pl. VIII.31); d–e) Tell el-Maskhuta (after Holladay 1997, pl. 7.22E); f) Sedment (after Petrie and Brunton 1924, pl. LXIII.1723a).

is of the ring-and-dot variety. At Sawama similar inlays appeared in only 4 of the 161 tombs excavated, a total of 2.5% (Bourriau and Millard 1971). These figures seem quite low when compared to the material from Jericho, where inlays appear in 71% of tombs within this date range (Kenyon 1960; 1965). They suggest that bone inlays of this kind were much more popular in Palestine than in Egypt, and it is possible that when they appear in Egypt they are reflecting Canaanite tastes. Yet is this really the case? Other small finds from the cemeteries at both Esna and Sawama show little Canaanite influence: in particular, classes of objects such as Canaanite weapon types, toggle pins or amulets, which might all be suggested to have a strong link to foreign modes of dress or religious symbolism, seem to be lacking. This suggests that these boxes with their inlays could have arrived at the site through simple trade, and may not have been carried or used there by

Canaanites. It may be significant that inlays also do not appear to be a common component of funerary assemblages in Eastern Delta sites such as Tell el-Dab'a, Tell el-Maskhuta or Tell el-Yehudiyeh, where other markers indicate a strong cultural link to Palestine.[1] None of this precludes these inlays having been produced by Canaanite craftsmen in either Palestine or Egypt. However it does suggest that they should be seen as a less important ethnic marker than some other artefact types.

RELIGION

As contact between Egypt and the Levant increased, Asiatic deities came to be worshipped in Egypt. This process appears to have begun in the period of Hyksos influence in the Delta, becoming more pronounced from the end of the 18th Dynasty. Temples to Canaanite gods such as Reshef, Qadesh, Anat and Astarte were established at Memphis, Giza, Pi-Ramesses, The Fayum and Deir el-Medineh, with some deities developing their own priesthoods (Kitchen 1969, 89; David 1986, 83). Several Egyptian stelae to Canaanite gods are also known (Pritchard 1969, 250; Jørgensen 1998, Figs. 39, 120; Petrie Museum UC 14392, 14400–1), while Canaanite deities were sometimes used as compound elements in Egyptian personal names, with names such as Anath-em-nekhu, Astart-em-heb and Baal-khepeshef becoming popular in the New Kingdom (Kitchen 1969, 87-90; Drower 1973, 482–3). To what degree the worship of these gods was confined to those of Canaanite descent is not clear. There appears to have been some syncretism, with Canaanite deities becoming the counterparts of Egyptian gods or goddesses with similar functions. Thus Reshef became identified with the Egyptian god Herishef, the 'Lady of Byblos' with Hathor, or Ba'al with Seth (Martin 1999, 204; Espinel 2002, 118; Pritchard 1969, 249–50; Redford 1993, 117–8).

In many of the above cases, Canaanite cult appears to have been practised in a largely Egyptian manner, and the kinds of cult installations and ritual objects known from the Levant seem to be missing, such as foreign types of libation tables, basins and cult stands (Kitchen 1969, 90). However, there are a few exceptions where the manner of worship appears to have been as foreign as its target. One such example is provided by a temple of Canaanite plan discovered at Tell el-Dab'a, along with various buried offering deposits of animal bones (Bietak 1996, 40–1; van den Brink 1982, 6; Müller 1998, 803).

Another, more personal form of cult activity is represented by foreign amulet types. As symbols representing a particular belief system, these are likely to reflect the cultural background of their owner. One popular type of Canaanite amulet is characterised by a crescent-shaped body (McGovern 1985, Type IV.B.1, 68). These are found throughout the Near East, usually in precious metals, and range in date from the Middle Bronze Age through to Iron Age I (McGovern 1985, 68, 70). Silver, glass and faience examples appear occasionally at New Kingdom sites such as Tell el-Amarna, Zawiyet el-Aryan, Gurob and Hemamieh (Figures 3.6a–b; Frankfort and Pendelebury 1933, Pl. XLIII.4; Dunham 1978, Pl. XXXIX 3rd row,

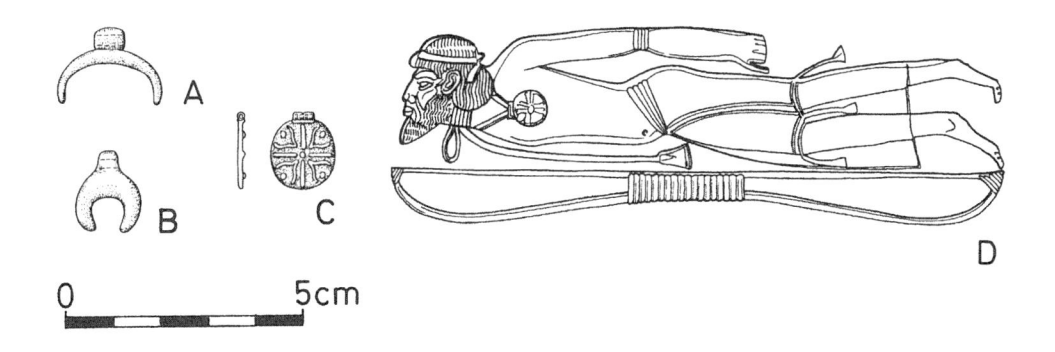

Figure 3.6 Canaanite amulets. a) Gurob (after Brunton & Engelbach 1927, pl. XLIII.44Z); b) Hemamieh (after Brunton, Qau and Badari III, pl. XXXII.17; c) Gurob (after Brunton & Engelbach 1927, pl. XLIII.44X; d) Detail from a footstool, Tomb of Tutankhamen (after Desroches Noblecourt 1963, pl. XI top).

far left; Brunton and Engelbach 1927, Pl. XLIII.44Z; Brunton 1930, Pl. XXXII.17). Another type of Canaanite amulet features a circular disc decorated with rays (McGovern 1985, Type VI.G.1–2). These appear in sheet metal, faience and glass, have a similar distribution pattern to the crescent-shaped variety, but are confined to Late Bronze Age deposits (McGovern 1985, 75–77; for this amulet in figurative art, see Woolley 1955, Pl. LVId AT/37/772). Egyptian art occasionally depicts Asiatics wearing this kind of amulet around their necks (Figure 3.6d; Davies and Faulkner 1947, Pl. VIII). A glass version was found in a New Kingdom context at Gurob, while a gold example currently in the Petrie Museum is said to be from Harageh (Figure 3.6c; Brunton and Engelbach 1927, Pl. XLIII.44X; Petrie Museum UC 6393). For discussions on the various Canaanite and Mesopotamian deities thought to be represented by both types of amulets, see James and McGovern 1993, 151 and Maxwell-Hyslop 1971, 140–151.

BURIAL CUSTOMS

Funerary customs are one aspect of cultural behaviour that may be a means of maintaining and transmitting group identity. There is insufficient space to do this topic justice here, but a few points should be made with relation to burial practices and Canaanite ethnicity. There is considerable variation in burial types available in the Levant during the course of the Bronze Age (for studies addressing this, see Abercrombie 1979; Stiebing 1970; Gonen 1992; Hallote 1994; Braunstein 1998). The degree to which the physical form of a burial is related to cultural rather than economic or topographical factors is debatable. The preference for rock-cut chamber tombs at Jericho, for example, may reflect no more than the number of such tombs cut in the earlier Early Bronze period which were available for reuse.

A different landscape means that comparable rock-cut tombs do not appear in coastal regions such as the Sharon Plain or Egyptian Delta. The cultural significance of this form therefore lies not so much in the physical shape of the tomb, but in the underlying concept, of a 'family' vault that could be reused over a period of time. This is similar to the idea behind the tombs built under house and courtyard floors for the inhabitants of sites such as Ugarit or Megiddo, although the execution differs.

These considerations suggest that some aspects of burial practice may well change when a group moves to a new geographical location. Those elements that do *not* change will be those that the group deem to be ethnically or culturally significant, and these are the features that we should look for when trying to locate foreigners in an Egyptian setting. Customs that may define Canaanite, as opposed to Egyptian funerary ritual include the relationship between tombs and living space, evidence for ritual meals and libations, the way in which the body itself is laid out within the tomb, and special associated features such as equid burials. At present, these kinds of customs seem to be largely confined to sites in the Eastern Delta, such as Tell el-Dab'a, Tell el-Maskhuta, and Tell el-Farasha (van den Brink 1982; Wapnish 1997, 360). The tombs here also display a variety of Canaanite object types, such as toggle pins, foil headbands, earrings, and weaponry along with both local and imported ceramics (Bietak 1991; Holladay 1997). These are combined with Egyptian objects such as mirrors, stands and harpoons, suggesting a mix of traditions (Philip 1995).

THE LEGACY OF CANAAN ON EGYPTIAN CRAFT AND TECHNOLOGY

This survey shows that there were a number of objects present in Egypt that were either imported from the Levant or produced locally to Canaanite designs, and it seems probable that at least some of this material was directed at Canaanites resident in Egypt. The growing numbers of these foreign residents must have strengthened the market for these kinds of products, and it may be hypothesised that it was this kind of market pressure that led to the initial adoption of many imports that went on to become locally produced Egyptian staples, such as 'Canaanite' amphorae or Tell el-Yehudiyeh ware juglets and their products.

Through such means, as well as through taxation, military booty and diplomatic gift exchange, Canaanite styles of art, iconography and form became more commonplace in Egypt, and a potential template for producing imitative and derivative styles in Egyptian workshops. The impact of foreign products on Egyptian workshops must have been increased by the growth in numbers of Canaanite artisans living and working in Egypt. Many of these resident foreigners brought with them skills in a variety of crafts and trades, skills that would have been applied, at least in part, to producing goods for the Egyptian consumer. More importantly, their presence made it possible to introduce technological as well as purely stylistic innovations. The influence of foreign artisans may be detected in part through the introduction of Semitic loanwords into the Egyptian vocabulary, especially where those terms relate to new crafts and techniques (Kemp and Vogelsang-Eastwood 2001, 55; Kitchen 1969, 84; Moorey 2001, 6; Redford 1993, 214–5; Shaw 2001, 65). The following sections demonstrate some areas where this influence may be identified through archaeological remains.

Textiles

Egyptian methods of yarn and cloth production are known from a variety of sources, including tomb paintings and funerary models depicting workshop scenes (Kemp and Vogelsang-Eastwood 2001, 68–9, 77–9, 317–29, 335–38). It is apparent from both these and archaeological remains that there were changes in weaving technology between the Middle Kingdom and the New Kingdom period. However an absence of clear sources for the interim period have made it difficult to document this transitional phase, and hence determine the mechanisms by which change was effected (Kemp and Vogelsang-Eastwood 2001, 310, 404). Nonetheless, there are a number of reasons why it seems likely that Canaanites were involved in this process.

Firstly, textile production is one of the crafts in Egypt that used a high proportion of Asiatic personnel, often acquired as prisoners of war (*e.g.*, Hayes 1955, 90; Kemp and Vogelsang-Eastwood 2001, 453). A wooden heddle jack from Lahun bore a proto-Canaanite inscription, suggesting just such an owner; this can be dated to either the 16th or 14th centuries B.C. (Cartwright *et al.* 1998, 92; Dijkstra 1990, 55–6). While the presence of Canaanite textile workers may not explain the full process of technological change, it does offer a plausible environment in which new ideas and methods might be introduced. Secondly, some of the innovations that can be demonstrated to occur in the textile industry between the 12th Dynasty and 18th Dynasty reflect practices already in use in the Levant during the Middle Bronze Age. This lends support to the theory that these innovations originated somewhere in that region.

A new type of spindle whorl appears in Egypt at the close of the Middle Kingdom (Vogelsang-Eastwood 2000, 266, 272). This is characterised by a hemispherical or conical body, a shape already widespread throughout Palestine (Figures 3.7a–b; *e.g.*, Kenyon 1965, Fig. 102). This co-exists with the flat Egyptian whorl into the later 18th Dynasty, as suggested by examples from Tell el-Amarna (Kemp and Vogelsang-Eastwood 2001). Even if this type of whorl did originate in the Levant, it could have been subsequently adopted by additional groups. A more useful guide for detecting the activities of non-Egyptian spinners may therefore lie in the actual spinning *technique*. This would include where the whorl was positioned on the spindle, as Egyptians preferred to place it at the top of the shaft (Kemp and Vogelsang-Eastwood 2001, 266–7). While evidence for Canaanite preferences in this respect is limited, some whorls from Megiddo were found placed at the centre of their spindles (Guy and Engberg 1938, 170–1, Fig. 175.6).

The direction of spin used to produce the thread or yarn may also help identify where textile workers were trained. In the Bronze Age, linens tended to be spun anti-clockwise (S-spun), because of the natural tendency of the flax to twist in

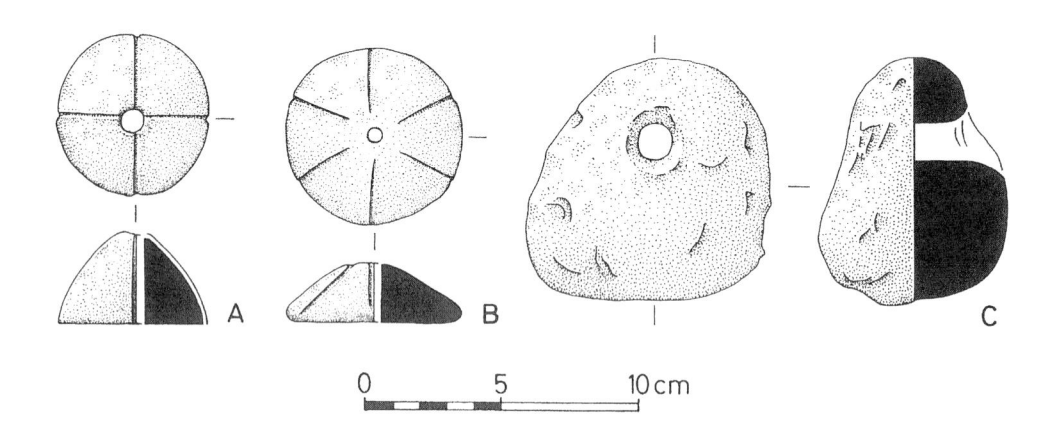

Figure 3.7 Spindle whorls and loomweights, Tell Heboua (a–c, after el-Maksoud 1998, figs 44.444–5, 43.439).

this direction, and this method was typical of Egyptian spinning practices in general (Kemp and Vogelsang-Eastwood 2001, 59). However wool produced in the Levant appears to have been spun in the opposite direction (Z-spun). As Z-spun fibres are not evident in any textiles found in early period Egypt, this implies that when the technology does appear in Egypt, it is something that had been introduced from the north, either through imported cloth or craftsmen (Sheffer and Tidhar 1988, 230). The nine examples of Z-spun yarn and single example of Z-spun wool found at Tell el-Amarna may therefore be the products of Canaanites working in the spinning industry there (Kemp and Vogelsang-Eastwood 2001, 59–60). Of course, it must be pointed out that the presence of fibres spun by foreigners need not have any implications for the nationality of those involved in weaving it into pieces of cloth, as these two activities were often conducted in separate locations (Kemp and Vogelsang-Eastwood 2001, 451; Friend 1998, 6).

The other major innovation was the introduction of the upright two-beamed loom, which was able to produce wider lengths of cloth than previously possible (McDowell 1986, 228, 236; Brovarski *et al*. 1982, 181, Fig. 47; Vogelsang-Eastwood 2000, 278). This loom first appears in Egyptian scenes early in the New Kingdom, coexisting with the traditional horizontal loom for the rest of the 2nd millennium B.C. It may be indicated archaeologically by the presence of pairs of stone socket blocks used to hold the looms in position, several of which were found in houses at Tell el-Amarna (Kemp and Vogelsang-Eastwood 2001, 60–1, 373; for possibly related blocks from Palestine, see Friend 1998, 10). Another loom type, the warp-weighted loom, was similarly vertical but with the addition of a series of hanging weights in stone or clay to provide the necessary tension for the hanging warp ends (Friend 1998, 4; Kemp and Vogelsang-Eastwood 2001, 392). It has been argued that the warp-weighted loom was also in use in New Kingdom Egypt, with loomweights appearing at sites such as El-Lisht, Kom Rabi'a, Tell el-Maskhuta and Tell el-Yehudiyeh (Figure 3.7c; Holladay 1997, Pl. 7.22C; Petrie 1906, Pl. XXIB.6–11; for a dissenting voice, see Kemp and Vogelsang-Eastwood 2001, 394). One major difficulty in determining how widespread this may have been lies in the fact that objects such as perforated weights have tended to be under-reported in archaeological publications.

Imported and foreign clothing types and styles are suggested by the occasional reference in Egyptian texts ('Asiatic *d3iw*-cloth', Warburton 1997, 159, n. 459), and by some of the textiles from New Kingdom royal burials, such as those found in the tomb of Tutankhamen (Hall 1986, 43–44; Lilyquist 1998, 215; Kemp and Vogelsang-Eastwood 2001, 445). Textiles with complex warp-patterned designs, sometimes known as compound weave, have been found primarily in New Kingdom royal tombs; the scarcity of the type and its élite associations have led to the suggestion that these textiles were imported or made by foreign weavers (Hall 1986, 45–6; Vogelsang-Eastwood 2000, 275; Kemp and Vogelsang-Eastwood 2001, 436). Another type of textile with hundreds of sewn-on gold rosettes, known from the Tomb of Tutankhamen, also seems to reflect foreign styles (Hall 1986, 40; possibly representing the loanword *ktt*, Kemp and Vogelsang-Eastwood 2001, 436–7). Textiles are known to have been taken as booty, sent with the dowries of foreign princesses, and presented as diplomatic greeting gifts, so this is entirely possible (Pritchard 1969, 238; Moran 1992, EA 22, EA 24; for this kind of exchange in the rest of the Near East, see Cochavi-Rainey 1999). However textiles were also sent *out* from Egypt, and these included coloured and decorated cloth that may have appealed to Asiatic tastes more than the plain white Egyptian linens (Moran 1992, EA 14: III.26–33; Cochavi-Rainey 1999, 197–9, 203, KbO 28 14, KUB 4 95, KbO 28 36, KbO I 29 + KbO IX 43, KUB 34 2).

It seems quite likely that some of the weavers who had come to Egypt from the Levant, whether as war booty or as an exchange of skilled personnel between courts, may have been set to work for the state producing these kind of garments, either for use by the royal household or as gifts to be sent abroad (Kemp and Vogelsang-Eastwood 2001, 476). It would have been in such workshops that more hybrid styles of garments were developed, such as the tunic of Tutankhamen that combines Syrian with Egyptian motifs (Hall 1986, 43–4; Moorey 2001, 7). Other 'Asiatic' elements appearing in the New Kingdom for the first time may include the use of coloured threads in cloth (Vogelsang-Eastwood 2000, 278). Dyeing cloth was not common Egyptian practice until then, in contrast to the elaborately patterned clothing usually associated with Canaanites in Egyptian art. Some of the dyes that begin to be used in the New Kingdom, such as indigotin and alizarin, are not native to Egypt and were probably introduced to the region from the Levant during the 18th Dynasty (Vogelsang-Eastwood 2000, 278–9).

Metalworking

Luxury metalwork was often targeted for theft or recycling, and consequently has a poor survival rate in the archaeological record. However, existing collections such as the jewellery from Middle Kingdom Dahshur or Lahun, or material from the tombs of the three wives of Tuthmosis III at Thebes suggest that there may be considerable foreign influence to be found in objects of this kind. Some of these objects bear the stylistic and technological signs of Canaanite origin, and were probably acquired through trade or warfare. Others have a more hybrid style, combining Asiatic elements with Egyptian iconography, suggesting local production by foreign craftsmen. This hybridisation might be greater where objects were manufactured by several specialists, only some of whom trained in Canaanite workshops.

This may well be the case for some of the richly ornamented gold objects in the tomb of Tutankhamen. These incorporate a number of purely Egyptian elements in their design, suggesting that the majority were made in Egyptian workshops rather than imported from abroad. However there are some features that point to the involvement of foreign craftsmen. These include motifs such as the running spiral, seen as a border on one of Tutankhamen's thrones, on a pair of elaborate earrings and some of the wooden staves (James 2000, 240, 268, 279, Fig. 2; Politis 2001, 184–6), or the style of the granulation work on some bracelets and sticks (James 2000, 246, 268, 279). Granulation is a technique that appears occasionally on objects from the Middle Kingdom, although it may not have been actually used by local craftsmen until the end of the Second Intermediate Period; it was probably introduced from Western Asia (Bourriau 1988, 147–8; David 1986, 55; Politis 2001, 182–4). One of the sheathed gold daggers from the tomb mixes elements such as the running spiral, guilloche, and free field animal scenes, with hieroglyphic inscriptions and Egyptian cloisonné work (Desroches-Noblecourt 1963, Pl. XXIA-B).

Another example of Canaanite crafts being applied in an Egyptian setting is represented by a gold and silver jug that was found at Tell Basta with objects ranging in date from Ramesses II to Tawoseret (Freed 1988, 148–9). The style of this vessel is a mix of both Egyptian and Asiatic elements. It had a handle in the shape of a rampant goat, with the body surface decorated with a number of registers. The upper part of the neck depicts a griffin and other animals in combat, while below are a series of hunting and fighting scenes. The main body is decorated with the figure of a royal butler, Atumemtaneb, who is depicted wearing Egyptian dress but worshipping a Canaanite goddess (Freed 1988, 148–9, Cat. 17). In the Ramesside period a number of Canaanites rose to prominence in the Egyptian palace administration, and the position of 'royal butler' was a popular career path in this respect (Redford 1993, 224; Schulman 1990, 20). It is possible that in Atumemtaneb we see one such figure, a man of Canaanite origins who chooses to portray himself in appearance and name as Egyptian, but who still followed the religion of his ancestors and retained the taste for some Canaanite style goods.

Ivory and Woodworking

Asiatic skills in working bone and ivory are suggested by some of the products found in the Late Bronze Age palace and city at Ugarit, and the vibrant, often 'egyptianising' decorative box and furniture inlays that are one of the great features of Near Eastern art of the period. Less is known about their skills in working wood, although the repertoire of tables, stools and beds found at Jericho, along with minor objects such as a range of boxes, vessels and platters, suggest that its practitioners were equally skilled. Indeed, the techniques and toolkits required for working all these materials would have been similar, and developments in one medium could be expected to be mirrored in comparable materials.

There are a few types of object known from Egyptian contexts that suggest Canaanite woodwork was known outside of the Levant, although it is difficult to determine whether these are simple imports, or objects produced in the region by Canaanite immigrants. Some types of carving, for example, do not seem to be particularly Egyptian, and may well be introduced from the Levant when they do occasionally appear, such as low relief work and openwork designs (Krzyszkowska and Morkot 2000, 325; Liebowitz 1977, 97). Wooden bowls with four ram's-head handles and incised linear or ring and dot designs around the rim are known from Lahun and Sedment (Petrie 1890, Pl. VIII.3; 1891, 11; Petrie and Brunton 1924, Pl. XLI.23). The type is much better represented at Jericho, where copies were also made from local gypsum (Sparks 1991). Another introduced form is the carinated pyxis with lug handles. This occurs in ivory, faience and gypsum workshops in the Levant, with the occasional example appearing in Egypt (Sparks 2001, 104, Fig. 6.6C–D and references *infra*; von Bissing 1904, Pl. VII.18742a–b, Vandier d'Abbadie, Cats. 123–127). One 18th Dynasty example from Gurob combines both Canaanite and Egyptian elements, with a carved base in the shape of a rosette, sphinx motifs, and lug handles that have been transformed into Hathor heads (Vandier d'Abbadie 1972, 47, Cat. 126). At the upper end of the range produced by foreign artisans is a ebony senet game from Dra Abu'l Naga, in Thebes, dating to the 17th Dynasty. This is decorated with a series of ivory inlays, combining an Egyptian style sphinx with an Asiatic tree of life flanked by ibexes, one of which is missing (Freed 1988, 207, Cat. 72). It is precisely this kind of mixture of styles that suggests the involvement of foreign artisans in an Egyptian setting.

MECHANISMS FOR CULTURAL OR TECHNOLOGICAL EXCHANGE

It now remains to consider the various mechanisms by which Canaanites came to live and work within Egypt, as the means by which they arrived, and their status in the communities they entered will have an impact on the nature of the physical remains they left behind them. One common source of foreign workers was through warfare. Men, women and children are often listed amongst booty sent back to Egypt after military campaigns in the Levant. The numbers involved could be considerable: it has been estimated that round 5,000 people were taken in this way between Years 23–42 of Tuthmosis III alone (Weinstein 1981, 14; Ahituv 1978, 103). Trade in slaves also took place outside the context of warfare, with Canaanite vassals selling people on to Egypt, while a letter in the archives at

Ugarit mentions an Egyptian slave dealer operating in the region (Redford 1993, 221; Drower 1973, 502). It may be more difficult to detect the presence of such groups in the Egyptian community, because as dependants of their masters, whether individuals, temples or royal estates, they were likely to be fed, equipped and clothed by them, and hence their material culture may well reflect Egyptian rather than foreign practices. The exception might be if they were expected to carry out skilled activities such as weaving or woodworking, where they might arrive with the tools of their trade, and the technology and methods applied may well reflect their previous training and experience.

There were also a number of opportunities for Canaanites to enter Egyptian society of their own free will. One popular occupation that seems to have attracted foreign personnel was the army. Asiatic mercenaries appear in Egyptian tomb scenes at Beni Hasan, equipped with foreign weapon types such as the duckbill or fenestrated axe (Newberry 1893, Pls. XVI, XXI, XLVII). Archaeological correlates for the use of Canaanite weapons may be provided by the assemblages found in 'warrior burials' at sites such as Tell el-Dab'a and Tell el-Maskhuta in the Eastern Delta (Phillip 1995; Bietak 1996, 36). There are further hints of the use of some of these axe types in the shape of the wounds that killed the 17th Dynasty pharaoh Seqenenre (Redford 1993, 126, Pl. 12); a woman and dog buried at Tell el-Maskhuta are also said to have both been dispatched with an Asiatic-style shaft-hole axe (Holladay 1982, 45). In the New Kingdom, representations of Canaanites serving in the Egyptian army as soldiers or royal bodyguards continue (*e.g.*, Berlin 14122, see below; Redford 1988, 15–17). These visual images are complemented by military positions such as the 'officer in command of Asiatic troops,' known from texts at sites such as Lahun and in the Western Desert (David 1986, 190; Himelfarb 2000, 21). If Egyptian art is correct in showing foreign soldiers retaining distinctive weapon types, we might expect to see their presence also reflected archaeologically through their deposition in tombs and houses during the New Kingdom.

Other occupations that seem to have led to Canaanites entering Egypt, whether as bound servants or free agents, include specialist professions such as craftsmen, musicians and merchants (Redford 1988, 18; Moorey 2001, 9–11). Diplomats and messengers also travelled into the country as representatives of Canaanite rulers and, like skilled personnel, could sometimes be detained on foreign soil for considerable periods. This was something that often gave cause for complaint (*e.g.*, Moran 1992, EA 3, EA 28). From the reign of Tuthmosis III, members of the royal family of vassal and allied Asiatic states were taken into the Egyptian court for political reasons. This included foreign princesses who were brought into the royal harem as a means of cementing political alliances. Texts such as the Amarna letters give us an idea of the types of objects they brought with them as part of their dowry. They were also accompanied by large numbers of servants; thus the Mitannian king Tushratta sent 270 women and 30 men to the Egyptian court when Tadukheba arrived to marry Amenhotep III (Moran 1992, EA 25: 64; Cochavi-Rainey 1999, 53).

The same period also saw the introduction of another policy that was equally influential in bringing Egyptian and Canaanite culture into close proximity: the practice of taking the chief heirs of Canaanite vassal princes back to Egypt as hostages, where they were raised in the Egyptian court until sent back to Canaan to take their father's throne (Moran 1992, EA 296: 23–29). In all these cases, the individuals concerned probably arrived with a variety of Canaanite objects in their possession, whether they were intending to stay for a short time or permanently. They would also provide a potential market for Canaanite goods for the duration of their stay. Over time, however, as exposure to Egyptian culture and practices increased, they may have adopted a more Egyptian lifestyle. Their eventual return could then become a mechanism for transporting Egyptian ideas and practices back into the Levant, leading to the growing 'Egyptianisation' of the upper classes there.

DISCOVERING ETHNICITY: WHAT MAKES A CANAANITE DIFFERENT?

This discussion is intended as a preliminary study of the Canaanite material available in Egypt during the 2nd millennium B.C. It demonstrates that there are numerous object classes of Canaanite style and origin to be found there at this time, including personal possessions, objects needed to carry out business and other craft activities, and a variety of products that show interaction between Canaanite and Egyptian artisans. However these items are probably only a fraction of those available, with many more hidden in museum storerooms and excavation archives, waiting to be recognised.

The most significant types of individual object in terms of identifying resident Canaanites would seem to be those which had a distinct form or function that was not paralleled in Egyptian material culture. Thus types such as the toggle pin, sheet metal 'headbands' and appliqués or drinking-straw strainers are probably more useful cultural indicators than objects such as bracelets, decorative box inlays or combs. The latter may exhibit a style that is foreign, but as similar items already had a role in Egyptian households it will be difficult to demonstrate that individual examples belonged to those from a Canaanite background. Any single object can be removed from its cultural setting by someone who valued it for some other aspect of its character, such as the constituent material, or as a keepsake representative of a visit abroad. In order to be more confident of assigning 'cultural' significance to how an artefact was used, it needs to be found in association with other culturally charged artefact types. If these objects are also found in a setting that points to foreign systems of behaviour – such as distinct dietary, lifestyle or cult practices, so much the better. It is this total 'package' that ultimately provides a convincing argument for detecting resident foreigners in the midst of Egyptian society, a package that has so far been discovered in full only in the Eastern Delta between the late Middle Kingdom and early 18th Dynasty.

However, it should be pointed out that there will always be parts of this group that will remain fairly difficult to detect archaeologically. There will be those who choose to abandon Canaanite customs in order to blend into Egyptian society. It will be equally difficult to detect those who are of a low social and economic status, and thus will have fewer material possessions with which to express their ethnic identity. The tombs from Tell Heboua in the Eastern Delta are a case in point. Of the 73 burials published from this site, some 59 burials had no associated finds at all, with the remaining tombs possessing only a handful of finds each (El-Maksoud 1998). Finally, those who were isolated and lacked the support of a Canaanite community may not have possessed the means to pursue a foreign lifestyle, or to choose the manner in which they were ultimately buried.

It may also not be appropriate to look for a single pattern or group of characteristics to explain 'Canaanite' behaviour or cultural identity. On the one hand, while some of the artefact classes described in this paper may be classified as 'foreign' to Egypt, their boundaries in the Near East are fluid and characterise other Asiatic groups besides the Canaanites. On the other hand, other objects may be more *limited* in their distribution, and relevant to only some of the people we might describe generically as Canaanite. Although there is a tendency to try and 'normalise' archaeological assemblages across a wide area, if only to clarify chronological and cultural developments, regional variations have been recognised on a number of levels, including styles of pottery, minor arts such as ivory working and burial customs (Price-Williams 1975; Liebowitz 1977, 89 n. 3; Gonen 1985). Often it appears that broader geographical or cultural units are applied to the area by outsiders, to whom distinctions of geography or cultural groups seem less significant. Thus while others may perceive of Canaanites as a single, ethnic group, internal perceptions may have been somewhat different, and these differences may find expression in the patterns visible in the archaeological record.

A gap may also exist between the Egyptian ideal of how a Canaanite should appear, and the reality of how individuals presented themselves. Ethnicity should be viewed as a flexible concept, very much in the eye of the beholder, rather than a static fixed principle of identity. It is something that probably changed over time. As resident Canaanites began to assimilate with Egyptian culture, it seems likely that many of their foreign aspects of dress and hairstyle would have disappeared, and it may become increasingly difficult to isolate their descendents in Egypt through their material culture. Thus while their 'foreign' identity might be preserved in aspects of language, such as people's names, or the epithets applied to them by Egyptians, the visual correlate of this and our ability to detect it archaeologically may become lost.

ACKNOWLEDGEMENTS

The author would like to thank Graham Reed for the illustrations used in this article.

NOTES

[1] None of the published burials from Tell el-Dab'a contain inlays. Bietak (1996, 36) suggested that, while many of the males buried at Tell el-Dab'a between the late Middle Kingdom and Second Intermediate Period were foreign, the women were predominantly Egyptian. The apparent absence of Canaanite inlays might make sense if their use was gender-specific; however this has yet to be demonstrated.

REFERENCES

Abercrombie, J. R., 1979, *Palestinian Burial Practices from 1200 to 600 BC*. Unpublished Ph.D. thesis, University of Pennsylvania.

Ahituv, S., 1978, Economic Factors in the Egyptian Conquest of Canaan, *Israel Exploration Journal* 28, 93–105.

Aldred, C., 1970, The Foreign Gifts Offered to Pharaoh, *Journal of Egyptian Archaeology* 56, 105–116.

Amiet, P., 1992, *Sceaux-cylindres en hématite and pierres diverses, corpus des cylindres de Ras Shamra-Ougarit II*. Éditions Recherche sur les Civilisations, Paris.

Amiran, R., 1977, The Ivory Inlays from the Tomb at El-Jisr Reconsidered, *Israel Museum News*, 12, 65–9.

Arnst, C.-B., Finneiser, K., Müller, I., Kischlewitz, H., Priese, K.-H. and Poethke, G., 1991, *Staatliche museen zu Berlin: Ägyptisches Museum und papyrussammlung*. Phillip von Zabern, Mainz.

Aston, B. G., 1994, *Ancient Egyptian Stone Vessels: Materials and Forms*, Studien zur Archäologie und Geschichte Altägyptens 5, Hiedelberger Orientverlag, Hiedelberg.

Aston, D. A., 1996, *Egyptian Pottery of the Late New Kingdom and Third Intermediate Period (Twelfth-Seventh Centuries BC)*, Studien zur Archäologie und Geschichte Altägyptens 13, Heidelberger Orientverlag, Heidelberg.

Ayrton, E. R., Cunelly, C. T. and Weigall, A. E. P., 1904, *Abydos III*, Egypt Exploration Fund, London.

Bietak, M., 1991, *Tell el-Dab'a 5, Ein Friedhofsbezirk der mittleren Bronzezeitkultur mit Totemtempel und Siedlungsschichten*. Verlag der Österreichischen Akademie der Wissenschaften, Wien.

Bietak, M., 1992, An Iron Age Four-Room House in Ramesside Egypt, *Eretz-Israel*, 23, 10*–12*.

Bietak, M., 1996, *Avaris, the Capital of the Hyksos. Recent Excavations at Tell el Dab'a*. British Museum Press, London.

Bourriau, J., 1988, *Pharaohs and Mortals. Egyptian Art in the Middle Kingdom*. Cambridge University Press, Cambridge.

Bourriau, J. and Millard, A., 1971, The Excavation of Sawama in 1914 by G.A. Wainwright and T. Whittemore, *Journal of Egyptian Archaeology*, 57, 28–57.

Braunstein, S., 1998, *The Dynamics of Power in an Age of Transition: An Analysis of the Mortuary Remains of Tell el-Far'ah (South) in the Late Bronze and Early Iron Ages*. Unpublished Ph.D. dissertation, Columbia University.

Brovarski, E., Doll, S.K. and Freed, R.E. 1982. *Egypt's Golden Age: The Art of Living in the New Kingdom 1558–1085*, Museum of Fine Arts, Boston.

Brunton, G., 1930, *Qau and Badari III*. British School of Archaeology in Egypt, London.

Brunton, G. and Engelbach, R., 1927, *Gurob*. British School of Archaeology in Egypt, London.

Cartwright, C., Granger-Taylor, H. and Quirke, S., 1998, Lahun Textile Evidence in London, in S. Quirke (ed.), *Lahun Studies*, 92–5. SIA Publishing, Reigate.

Cochavi-Rainey, Z., 1999, *Royal Gifts in the Late Bronze Age: Fourteenth to Thirteenth Centuries B.C.E.* Beer-Sheva Vol. 13, Ben-Guryon University of the Negev Press, Beersheba.

Cohen, S., 2002, *Canaanites, Chronologies, and Connections: the Relationship of Middle Bronze IIA Canaan to Middle Kingdom Egypt*. Eisenbrauns, Winona Lake.

Collon, D., 1982, *The Alalakh Cylinder Seals*, BAR International Series 132. Oxford.

David, R., 1986, *The Pyramid Builders of Ancient Egypt. A Modern Investigation of Pharaoh's Workforce*. Routledge, London.

Davies, N. de G. and Faulkner, R. O., 1947, A Syrian Trading Venture to Egypt, *Journal of Egyptian Archaeology*, 33, 40–6.

Desroches-Noblecourt, C., 1963, *Tutankhamen*. Michael Joseph Ltd, London.

Dijkstra, M., 1990, The So-called 'Ahitub-Inscription from Kahun (Egypt)', Zeitschrift des deutsche Palästina-Vereins, 106, 51–6.

Downes, D., 1974, *The Excavations at Esna 1905–6*. Aris and Phillips, Warminster.

Drower, M., 1973, Syria c. 1550–1400 B.C., in *Cambridge Ancient History, Volume 2, part 1, History of the Middle East and Aegean Region, c.1800–1380 B.C*, 3rd edn., 417–525. Cambridge University Press, London.

Dunham, D., 1978, *Zawiyet el-Aryan: The Cemeteries Adjacent to the Layer Pyramid*. Museum of Fine Arts, Boston.

El-Maksoud, M. A., 1998, *Tell Heboua (1981–1991): enquête archéologique sur la deuxième période intermédiaire et le nouvel empire a l'extrémité orientale du Delta*. Éditions Recherche sur les civilisations, Paris.

Espinel, A. D., 2002, The Role of the Temple of Ba'al at Gebal as Intermediary Between Egypt and Byblos During the Old Kingdom, *Studien zur altägyptischen Kultur*, 30, 103–119.

Frankfort, H., 1927, *Studies in Early Pottery of the Near East*. Royal Anthropological Institute of Great Britain and Ireland, London.

Frankfort, H. and Pendlebury, J. D. S., 1933, *The City of Akhenaten II: the North Suburb and the Desert Altars*. Egypt Exploration Fund, London.

Freed, R., 1988, *Ramses the Great*. Boston Museum of Science, Boston.

Friend, G., 1998, *Tell Taannek 1963–1968 III: The Artifacts, 2: the Loomweights*. Palestinian Institute of Archaeology, Birzeit University.

Germer, R., 1998, The Plant Remains found by Petrie at Lahun and Some Remarks on the Problems of Identifying Egyptian Plant Names, in: S. Quirke (ed.), *Lahun Studies*, 84–91, SIA Publishing, Reigate.

Giddy, L., 1999, *Kom Rabi'a: the New Kingdom and Post-New Kingdom objects*. Egypt Exploration Society, London.

Gonen, R., 1985, Regional Patterns of Burial Customs in Late Bronze Age Canaan, *Bulletin of the Anglo-Israel Archaeological Society*, 4, 1984–5, 70–4.

Gonen, R., 1992, *Burial Patterns and Cultural Diversity in Late Bronze Age Canaan*. Eisenbrauns, Winona Lake.

Griffith, F. Ll., 1890, *The Antiquities of Tell el Yahûdîyeh and Miscellaneous Work in Lower Egypt during the Years 1887–1888*. Seventh Memoir. Egypt Exploration Society, London.

Griffith, F. Ll., 1926, A Drinking Siphon from Tell el-'Amarnah, *JEA*, 12, 22–3.

Guy, P. L. O. and Engberg, R. M., 1938, *Megiddo Tombs*. University of Chicago Press, Chicago.

Hall, R., 1986, *Egyptian Textiles*. Shire Publications, Aylesbury.

Hallote, R., 1994, *Mortuary Practices and their Implications for Social Organization in the Middle Bronze Southern Levant*. Unpublished Ph.D. thesis, University of Chicago.

Hayes, W. C., 1955, *A Papyrus of the Late Middle Kingdom in the Brooklyn Museum: Papyrus Brooklyn 35.1446*. Brooklyn Museum, New York.

Henschel-Simon, E., 1937, The Toggle-Pins in the Palestine Archaeological Museum, *Quartrly of the Department of Antiquities, Palestine*, 6, 169–209.

Heltzer, M., 1994, Trade Between Egypt and Western Asia: New Metrological Evidence (on E. W. Castle in JESHO XXXV), *Journal of the Economic and Social History of the Orient*, 37, 318–321.

Higginbotham, C. R., 2000, *Egyptianization and Élite Emulation in Ramesside Palestine: Governance and Accomodation on the Imperial Periphery*. Brill, Boston.

Himelfarb, E. J., 2000, First Alphabet Found in Egypt, *Archaeology*, 53.1, 21.

Holladay, J. S., Jr., 1982, *Cities of the Delta Part III: Tell el-Maskhuta: Preliminary Report on the Wadi Tumilat Project 1978–1979*. Udena Publications, Malibu.

Holladay, J. S., Jr., 1997, The Eastern Nile Delta During the Hyksos and Pre-Hyksos Periods: Toward a Systemic/Socio-Economic Understanding", in E. D. Oren (ed.), *The Hyksos: New Historical and Archaeological Perspectives*, 183–252. The University Museum, University of Pennsylvania, Philadelphia.

James, F. W. and McGovern, P. E., 1993, *The Late Bronze Egyptian Garrison at Beth Shan: A Study of Levels VII and VIII*, The University Museum, University of Pennsylvania, Philadelphia.

James, T. G. H., 2000, *Tutankhamun: The Eternal Splendour of the Boy Pharaoh*. Tauris Parke Books, London and New York.

Jørgensen, M., 1998, *Catalogue Egypt II (1550–1080)*. NY Carslberg Glyptotek, Copenhagen.

Kemp, B. J. and Vogelsang-Eastwood, G., 2001, *The Ancient Textile Industry at Amarna*. Egypt Exploration Society, London.

Kenyon, K. M., 1960, *Excavations at Jericho I*. British School of Archaeology in Jerusalem, London.

Kenyon, K. M., 1965, *Excavations at Jericho II*. British School of Archaeology in Jerusalem, London.

Kitchen, K. A., 1969, Interrelations of Egypt and Syria, in M. Liverani (ed.), *La Siria nel Tardo Bronzo*. Centro per le antichita e la storia dell'arte del vicino oriente, Rome.

Krzyszkowski, O. and Morkot, R., 2000, Ivory and Related Materials, in P. T. Nicholson and I. Shaw, *Ancient Egyptian Materials and Technology*, 320–331. Cambridge University Press, Cambridge.

Leahy, A., 2000, Ethnic Diversity in Ancient Egypt, in J. M. Sasson (ed.), *Civilizations of the Ancient Near East volume I*, 225–234. Hendrickson, Peabody.

Liebowitz, H., 1977, Bone and Ivory Inlay from Syria and Palestine, *Israel Exploration Journal* 27, 89–97.

Liebowitz, H., 1986, Late Bronze II Ivory Work in Palestine: Evidence of a Cultural Highpoint, *Bulletin of the American Schools of Oriental Research*, 265, 3–24.

Lilyquist, C., 1998, The Tomb of Tuthmosis III's Foreign Wives: A Survey of its Architectural Type, Contents, and Foreign Connections, in C. J. Eyre (ed.)., *Proceedings of the Seventh International Congress of Egyptologists, Cambridge, 3–9 September 1995*. 677–681. Uitgeverij Peeters, Leuven.

Lilyquist, C., 1999, The Objects Mentioned in the Texts, in Z. Cochavi-Rainey, *Royal Gifts in the Late Bronze Age: Fourteenth to Thirteenth Centuries B.C.E.* Beer-Sheva Vol.13, 211–8. Ben-Guryon University of the Negev Press, Beersheba.

Mace, A. C., 1922, Loom Weights in Egypt, *Ancient Egypt* 1922, 75–6.

Martin, G. T., 1999, Alalakh 194: An Ancient Seal-Impression Re-interpreted, in: A. Leahy and J. Tait (eds.), *Studies on Ancient Egypt in Honour of H.S. Smith*, 201–207, Egypt Exploration Society, London.

Maxwell-Hyslop, R., 1971, *Western Asiatic Jewellery, c. 3000–612*. Methuen, London.

McDowell, J. A., 1986, Kahun: The Textile Evidence, in R. David (ed.), *The Pyramid Builders of Ancient Egypt*, Routledge, London and New York.

McGovern, P. E., 1985, *Late Bronze Palestinian Pendants. Innovation in a Cosmopolitan Age*, American Schools of Oriental Research Monograph 1, Sheffield.

McGovern, P. E. 1989, Cross-Cultural Craft Interaction: The Late Bronze Egyptian Garrison at Beth Shan, in P. E. McGovern (ed.), *Cross-Craft and Cross-Cultural Interactions in Ceramics*, 147–196, American Ceramic Society, Westerville.

Moorey, P. R. S., 2001, The Mobility of Artisans and Opportunities for Technology Transfer Between Western Asia and Egypt in the Late Bronze Age, in A. J. Shortland (ed.), *The Social Context of Technological Change: Egypt and the Near East, 1650–1150 BC*, 1–14. Oxbow, Oxford.

Moran, W. L., 1992, *The Amarna Letters*. John Hopkins University Press, Baltimore.

Mumford, G. D., 1998, *International relations between Egypt, Sinai, and Syria Palestine during the Late Bronze Age to Early Persian period (Dynasties 18 26: c.1550–525 B.C.): A spatial and temporal analysis of the distribution and proportions of Egyptian(izing) artefacts and pottery in Sinai and selected sites in Syria Palestine*. Unpublished Ph.D. thesis, University of Toronto.

Müller, V., 1998, Offering Deposits at Tell el-Dab'a, in C. J. Eyre (ed.), *Proceedings of the Seventh International Congress of Egyptologists*, 793–803. Uitgeverij Peeters, Leuven.

Negbi, O., 1976, *Canaanite Gods in Metal*. Kegan Paul, London.

Newby, P. H., 1980, *Warrior Pharaohs: The Rise and Fall of the Egyptian Empire*. Book Club Associates, London.

Oates, D., Oates, J. and McDonald, H., 1997, *Excavations at Tell Brak. Volume 1: The Mitanni and Old Babylonian Periods*. McDonald Institute for Archaeological Research, London.

Oren, E. D., 1987, The Ways of Horus in North Sinai, in A. Rainey (ed.), *Egypt, Israel, Sinai*, 69–119. Tel Aviv University, Jerusalem.

Petrie, W. M. F., 1890, *Kahun, Gurob and Hawara*. Kegan Paul, London.

Petrie, W. M. F., 1891, *Illahun, Kahun and Gurob*. Nutt, London.

Petrie, W. M. F., 1906, *Hyksos and Israélite Cities*. British School of Archaeology in Egypt, London.

Petrie, W. M. F., 1914, *Amulets*. British School of Archaeology in Egypt, London.

Petrie, W .M. F., 1926, *Ancient Weights and Measures*. British School of Archaeology in Egypt, London.

Petrie, W. M .F., 1931, *Ancient Gaza I*. British School of Archaeology in Egypt, London.

Petrie, W. M. F., 1934, *Ancient Gaza IV*. British School of Archaeology in Egypt, London.

Petrie, W. M. F. and Brunton, G., 1924, *Sedment I–II*. British School of Archaeology in Egypt, London.

Petrie, W. M .F., Mackay, E. J. H. and Murray, M. A., 1952, *City of Shepherd Kings and Ancient Gaza V*. British School of Egyptian Archaeology LXIV, London.

Philip, G., 1995, Tell el-Dab'a Metalwork: Patterns and Purpose, in: W. V. Davies and L. Schofield (eds.), *Egypt, the Aegean and the Levant. Interconnections in the Second Millennium B.C.*, 66–83. British Museum Press, London.

Politis, T., 2001, Gold and Granulation: Exploring the Social Implications of a Prestige

Technology in the Bronze Age Mediterranean, in: A. J. Shortland (ed.), *The Social Context of Technological Change: Egypt and the Near East, 1650–1150 BC*, 161–94. Oxbow, Oxford.

Price-Williams, D., 1975, *An Examination of MBAII typology and Sequence Dating in Palestine with Particular Reference to the Tombs of Jericho and Fara (South)*. Unpublished Ph.D. thesis, Institute of Archaeology University College London.

Pritchard, J. B., 1969, *Ancient Near Eastern Texts Relating to the Old Testament*. Princeton University Press, Princeton.

Pulak, C., 1997, The Uluburun Shipwreck, in: S. Swiny, R. Hohlfelder and H. Swiny (eds.), *Res Maritimae, Cyprus and the Eastern Mediterranean from Prehistory to Late Antiquity. Proceedings of the Second International Symposium, "Cities on the Sea," Nicosia, Cyprus, October 18–22, 1994*, 233–252. Scholars Press, Atlanta.

Quibell, M. J. E., 1901, A Tomb at Hawaret el-Gurob, *Annales du Service des Antiquités de l'Égypte*, 2, 141–3.

Randall-MacIver, D. and Mace, A. C., 1902, *El-Amrah and Abydos*. Egypt Exploration Society, London.

Redford, D., 1988, *The Akhenaten Temple Project Volume 2: Rwd-mnw, Foreigners and Inscriptions*. Akhenaten Temple Project, Toronto.

Redford, D. B., 1993, *Egypt, Canaan and Israel in Ancient Times*. Princeton University Press, Princeton.

Redmount, C. A., 1995, Pots and Peoples in the Egyptian Delta: Tell el-Maskhuta and the Hyksos, *Journal of Mediterranean Archaeology*, 8.2, 61–89.

Schaeffer, C. F. A., 1949, *Ugaritica II (Mission de Ras Shamra V)*. Librairie orientaliste Paul Geuthner, Paris.

Schaeffer, C. F. A., 1962, *Ugaritica IV*. Mission de Ras Shamra XV, Paris.

Shaw, I., 2001, Egyptians, Hyksos and Military Hardware: Causes, Effects or Catalysts?, in A. J. Shortland (ed.), *The Social Context of Technological Change: Egypt and the Near East, 1650–1150 BC*, 59–72. Oxbow, Oxford.

Schneider, T., 1992, *Asiatische Personennamen in ägyptischen Quellen des Neuen Reiches*. Orbis Biblicus et Orientalis 114, University Press Freibourg.

Schulman, A. R., 1990, The Royal Butler Ramessessami'on: an Addendum, *Chronique d'Égypte*, 65, 12–20.

Sheffer, A. and Tidhar, A., 1988, Identification of Textile Remnants, in B. Rothenberg (ed.), *The Egyptian Mining Temple at Timna*, 223–230. Institute for Archaeo-Metallurgical Studies, London.

Sparks, R. T., 1991, A Series of Middle Bronze Age Bowls with Ram's-Head Handles from the Jordan Valley, *Mediterranean Archaeology*, 4, 45–54.

Sparks, R. T., 2001, Stone Vessel Workshops in the Levant: Luxury Products of a Cosmopolitan Age, in A.J. Shortland (ed.), *The Social Context of Technological Change: Egypt and the Near East, 1650–1550 BC*, 93–112. Oxbow, Oxford.

Stiebing, W. H., Jr., 1970, *Burial Practices in Palestine During the Bronze Age*. Unpublished Ph.D dissertation, University of Pennsylvania.

Thomas, A., 1981, *Gurob: A New Kingdom Town*, Egyptology Today 5.1. Aris and Phillips, Warminster.

Tufnell, O., 1978, Graves at Tell el-Yehudiyeh – Reviewed after a Lifetime, in R. Moorey and P. Parr (eds.), *Archaeology in the Levant – Essays for Kathleen Kenyon*, 76–101. Aris and Phillips, Warminster.

Tufnell, O. and Ward, W., 1966, *Relations Between Byblos, Egypt and Mesopotamia at the End of the Third Millennium B.C.: A Study of the Montet Jar*. Geuthner, Paris.

van den Brink, E. C. M., 1982, *Tombs and Burial Customs at Tell el-Dab'a and their Cultural*

Relationship to Syria-Palestine during the Second Intermediate Period. Universität Wien, Vienna.

Vandier d'Abbadie, J., 1972, *Les objets de toilette égyptiens au Musée du Louvre.* Editions des Musées Nationaux, Paris.

Vogelsang-Eastwood, G., 2000, Textiles, in P. T. Nicholson and I. Shaw (eds.), *Ancient Egyptian Materials and Technology*, 268–298. Cambridge University Press, Cambridge.

von Bissing, F. W. 1904, *Catalogue général des antiquités Égyptiennes du Musée du Caire Nos 18065–18793: Steingefäße.* Adolf Holzhausen, Vienna.

Wapnish, P., 1997, Middle Bronze Equid Burials at Tell Jemmeh and Reexamination of a Purportedly "Hyksos" Practice, in: E. D. Oren (ed.), *The Hyksos: New Historical and Archaeological Perspectives*, 335–67. Philadelphia University Museum, Philadelphia.

Warburton, D. 1997, *State and Economy in Ancient Egypt: Fiscal Vocabulary of the New Kingdom.* Vandenhoeck and Ruprecht Gottingen, Fribourg.

Ward, W. A. 1994, Foreigners Living in the Village, in E. H. Lesko (ed.), *Pharaoh's Workers: The Villagers of Deir el Medina*, 61–85. Cornell University Press, Ithaca and London.

Weinstein, J. M. 1981, The Egyptian Empire in Palestine: A Reassessment, *Bulletin of the American Schools of Oriental Research*, 241, 1–28.

Woolley, C. L. 1955, *Alalakh – An Account of the Excavations at Tell Atchana in the Hatay, 1937–1949.* Oxford University Press, Oxford.

Ziffer, I., 1990, *At That Time the Canaanites Were in the Land: Daily Life in Canaan in the Middle Bronze Age 2, 2000–1550 B.C.E.,* Eretz Israel Museum, Tel Aviv.

Chapter 4

The Provenance of Canaanite Amphorae found at Memphis and Amarna in the New Kingdom: results 2000–2002

L. M. V. Smith, J. D. Bourriau, Y. Goren, M. J. Hughes and M. Serpico

Abstract

The provenance study reported on in the first Workshop is part of the Canaanite Amphorae Project, which examines the contexts, fabrics, contents and inscriptions on jars imported into New Kingdom Egypt. In the previous paper, the available sample of Canaanite Jars was divided into five major fabric groups, on the basis of study of the sherd chips by low-power binocular microscopy and thin-section petrography. We were able to posit regions of origin, on geological grounds, for these five fabric groups. The present paper will describe the current, and final, stage of the fabric and provenance study, incorporating findings made during the last two years. These include the recognition of a further fabric group, and some refinements to and, in one case, adjustment of the attributions to source regions made initially.

INTRODUCTION

The Canaanite Amphora Project examines the contexts, fabrics, and contents of these characteristic storage jars, which were imported into New Kingdom Egypt from the Syro-Palestinian region. The impetus for the project derives in part from an earlier study by Janine Bourriau, based on samples found in New Kingdom deposits at Memphis (Bourriau, 1990). She was able to distinguish several different fabrics and questioned whether these fabrics could correlate with different areas of production of the amphorae, which might be relatively restricted. Similar research at Amarna (Nicholson and Rose 1985, 139–40), along with research into the natural products stored in the amphorae (Serpico 1996; Serpico and White 2000) indicated that it might be possible to link specific fabrics to the transport of specific commodities. The ability to then correlate the areas of vessel production with the source areas of the contents could potentially link Egypt to important Eastern Mediterranean exchange networks during the Late Bronze Age.

The objectives of the project have been addressed by two major scientific studies: firstly, the investigation of the organic residues (Stern et al. 2000; Stern et al. 2003; Serpico et al. 2003) and, secondly, the study of the vessel fabric and vessel provenance[1] (Bourriau, Smith and Serpico 2001; Smith, Bourriau and Serpico 2000). Focussing here on the fabric aspect of the project, the methodology used for the current study generally followed that initiated by Bourriau and Nicholson (1992), which involved examination of the sherd break by low-power binocular microscopy and thin-section petrography. Full details of the procedure used for the Canaanite Amphora Project have now been published (Smith, Bourriau and Serpico, 2000). More recently, the thin-section petrography was complemented by a bulk chemical analysis by Inductively-coupled Plasma Atomic Emission Spectroscopy (ICP-AES),[2] to which was added chemical data from a previous analysis by Neutron Activation Analysis (Al-Dayel 1995). In the absence of comparative material from known production sites for the fabrics included in the present study, provenance was determined on the basis of comparison of sherds with local geology and with comparative sherds from sites of use, whose fabrics were considered to be representative of local geology.

INITIAL FINDINGS OF THE PROVENANCE STUDY

The present paper will begin with a summary of the state of the research as presented at the first Social Context of Technological Change Workshop, followed by an overview of the final results obtained and the attributions of provenance made during the intervening two years. At the time of this initial paper (Bourriau, Smith and Serpico 2001), the available sample of Canaanite Jars was divided into five major fabric groups, designated Groups 1–5, which each included a number of subgroups (see Figure 4.1). Further interim details of the provenance study and illustrations of the Fabric Groups and subgroups in sherd break and in thin section have also been published (Smith, Bourriau and Serpico 2000; Serpico et al. 2003).

Group 1

In this fabric, the non-plastic assemblage includes dense, well-sorted, sand-sized quartz and carbonate grains in varied proportions (Figure 4.2). Other components are chert and, occasionally, rounded grains of weathered olivine basalt or minerals derived from this rock (augite, olivine). The quartz sand is associated with a small proportion of other minerals, mainly feldspars, hornblende and zircon.

In general, the size of the quartz grains, their uniformity in size, and their degree of roundness and sphericity are characteristic of sediments that originate from the Nile River. These grains are deposited along the coastline of Sinai and

Group	Sub-group	Distinguishing characteristics
1	1.1	Abundant limestone and quartz, with a large difference in proportions
	1.2	Lower limestone content, abundant quartz with a small difference in proportions
	2.1	Low quartz and limestone content, with limestone more abundant than quartz
	2.2	Low quartz and limestone content, with the amount of limestone similar to that of quartz
	2.3	Low quartz and limestone content, with quartz more abundant than limestone
2	1.1	Low quartz content
	1.2	Medium quartz content
	2.1	High quartz content
3	1.1	Abundant volcanic rock fragments (>17%)
	1.2	Low content of volcanic rock fragments (c. 12%)
4	1.1 (High serpentine)	High ophiolite content, with high serpentine content
	1.1 (High radiolarian)	High ophiolite content, with high radiolarian chert content
	2.1	High ophiolite content, with abundant replacement chert and low radiolarian chert content
	2.2	Low ophiolite content, with low replacement chert content
5	1.1	High limestone content, with low microfossil content
	1.2	Lower limestone content, with low microfossil content
	2.1	High limestone content, with high microfossil content
	2.2	High limestone content, with high microfossil content: distinguished visually by 'Beige inclusions'

Figure 4.1 Summary of the petrographic Groups and Sub-groups, with features distinguishing between them in thin section and sherd break.

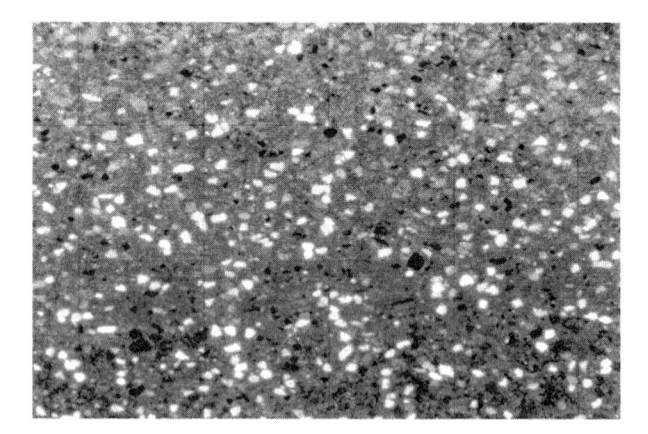

Figure 4.2 Example of Group 1 fabric. Specimen 76: showing abundant quartz and limestone (XPL: field width c. 10.7 mm).

Palestine and designated 'Coastal Sand' in the literature. The carbonates also include separate fossils of coralline algae, especially of the genus *Amphiroa*, also a characteristic of the coastal formations (Sivan 1996). The 'Coastal Sand' is combined with a small proportion of chert, chalk and basalt fragments; local additions supplied by rivers draining onto the Coastal Plain from the east.

The main area where 'Coastal Sand', basalt, chalk and chert could be found together is restricted to the seaward opening of the Jezreel Valley (Figure 4.3). This is where the Qishon river, draining the basaltic areas near Affula and elsewhere in the Jezreel Valley, and the Eocene formations to the north and south, finally reaches the region of 'Coastal Sand'. Hence, it was hypothesised that the most probable provenance of the first group of fabrics was the area of the Haifa Bay to the north of the Carmel Ridge, approximately between Haifa and Akko.

Group 2

This fabric group (Figure 4.4) is less homogeneous than Group 1, with the most striking difference being in the percentages of quartz, and the presence of coarse or very coarse rounded chalk inclusions. Some examples seem to be related to Group 1, in that they contained 'Coastal Sand' with calcareous inclusions and a small quantity of volcanic rock fragments. The degree of difference petrographically seemed sufficient to separate out this fabric as a specific Group, having a lower percentage of basaltic fragments than Group 1, and sometimes a highly silty and ferruginous matrix probably obtained from an iron-stained soil, locally termed *Hamra*.

Unfortunately, for the provenance study, this *Hamra* soil is spread along much of the Coastal Plain of Israel from the Ashdod area northwards. However, the Coastal Sand does not extend much beyond the Akko area on the northern coast of Israel. The relatively high (though variable) quartz content of Group 1, ranging up to c. 16%, seemed to rule out the northernmost part of the distribution of the Coastal Sands, which are largely composed of carbonates. From these lines of

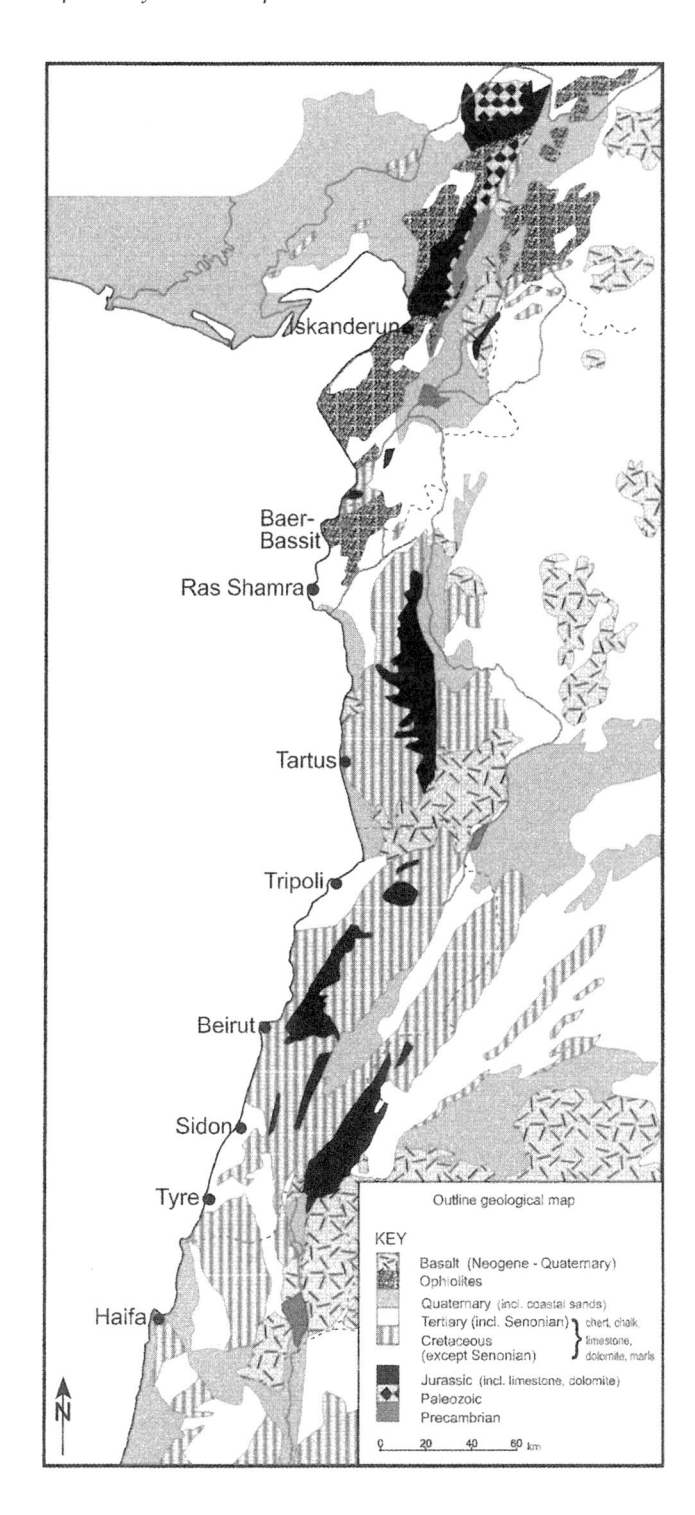

Figure 4.3 Simplified geological map of the region significant for vessel provenance. (After Beydoun 1977 and I.F.P-C.N.E.X.O 1974). (Map drawn by L. Smith using 'Professional Draw' package, by Gold Disk Inc.).

Figure 4.4 Example of Group 2 fabric. Specimen 80: with high quartz content and coarse chalk inclusions. (XPL: field width c. 10.7 mm).

Figure 4.5 Example of Group 3 fabric. Specimen 62: showing abundant volcanic rock fragments, predominantly olivine basalt. (XPL: field width c. 10.7 mm).

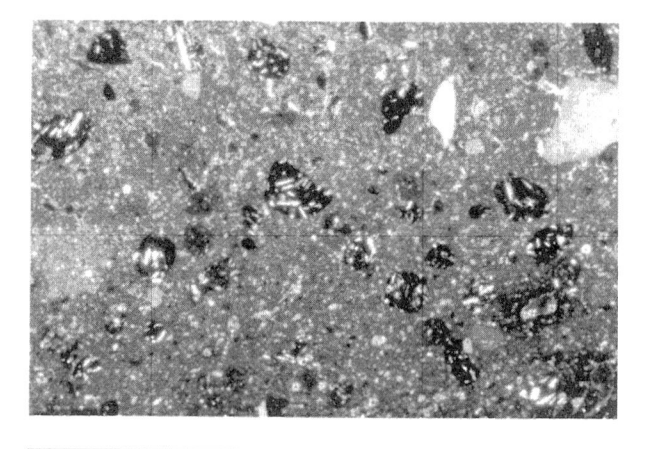

Figure 4.6 Example of Group 4 fabric. Specimen 187: with high proportion of the ophiolite-related inclusions, showing specimen with high percentages of serpentine. (XPL: field width c. 10.7 mm).

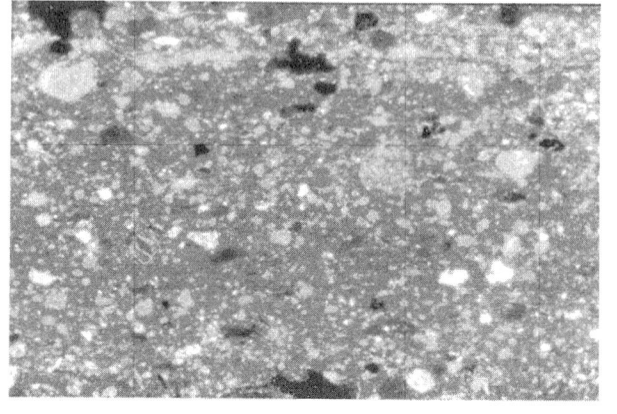

evidence, the source area for Group 2 could be limited only to the Coastal Plain of Israel, approximately between Ashdod and Akko. At that stage of the study, it did not seem possible to define the provenance more closely.

Group 3

Group 3 is characterised by abundant subangular to rounded basalt inclusions, combined with limestone, chalk and chert, and very low percentages of quartz above fine sand-size (Figure 4.5). The latter feature means that the quartz content of this Group is not comparable to the 'Coastal sand' of the previous two Groups. In sherd break and in thin section, the dark, medium to very coarse basaltic inclusions make this one of the most visually distinctive of the fabrics.

Initially, a source area was suggested through the combination of the absence of Coastal Sand and the presence of the basalts indicating an inland, not a coastal, origin. Moreover, the nature of the limestone and chalk inclusions could be interpreted as resulting from the incorporation of a wadi sand and basalt as temper. On this basis, the most likely origin appeared to be the central Jezreel Valley into eastern Galilee examples from several sites in this region, such as Megiddo, having relatively high content of basalt inclusions.

Group 4

In the case of Group 4, the characteristic features (Figure 4.6) included a combination of igneous rock fragments, including those showing alteration to iddingsite or serpentinite, together with quartzite, schist and both replacement and radiolarian chert. These indicated a source where components of an ophiolite complex exist; ophiolites being uplifted portions of oceanic crust containing a basic sequence of igneous rocks such as basalts and dolerite, together with metamorphic and sedimentary rocks, which include replacement and radiolarian chert.

The ophiolite-related constituents occur with pelagic and shallow water carbonates, including planktonic and benthic foramenifera. This combination is consistent with a possible origin in Turkey, north-west Syria or Cyprus. The most likely source of origin on geological grounds appeared to be the Baër-Bassit complex in northwest Syria. This was supported by examination of local pottery from the coastal site of Ras Shamra (Ugarit), immediately south-west of the Baër-Bassit, which contains the same ophiolite-related suite of minerals and lithic fragments with components indicating a coastal provenance. Further support came from the presence of a similar set of constituents in thin sections of two Amarna Letters known to have been sent from Ugarit, and analysed in a general study of the provenance of the Letters (Goren, Bunimovitz, Finkelstein and Na'aman, 2003).

Fabric Group 4 was a group for which we had suitable comparable specimens for chemical analysis to provide direct information on provenance. The implications of the chemical Discriminant Analysis for the provenance of this Fabric Group were most clearly seen in two runs, of which one is illustrated (Figure 4.7), in which all the comparative specimens from Ras Shamra were assigned to, and clustered well with, the Fabric Group 4 specimens. Conversely, both the Group 4 specimens and the Ras Shamra specimens clustered separately from the Comparative specimens considered characteristic of areas of Cyprus.

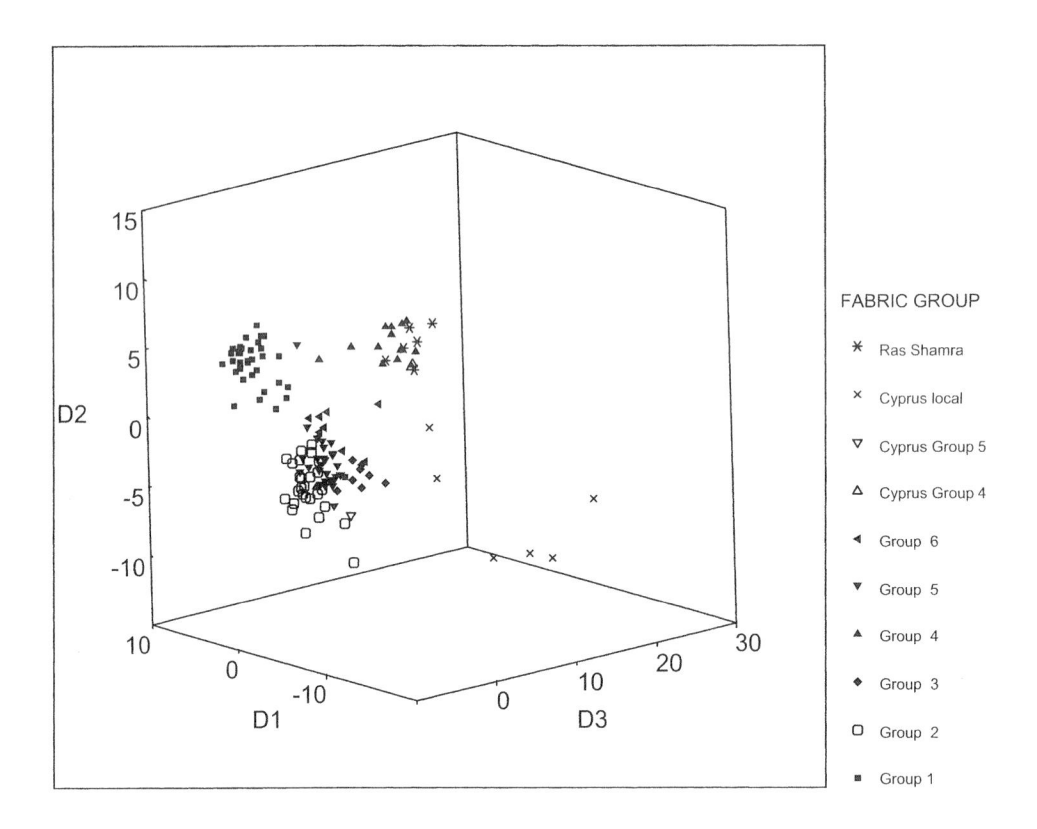

Figure 4.7 Plot from Discriminant Analysis of fabric groups on chemical data showing relationship between Group 4 and Ras Shamra specimens.

Hence, the Discriminant Analysis of the chemical data strongly supported the attribution of Group 4, on petrographic grounds, to the Baër-Bassit region.

Group 5

The paste in this Group is sometimes mixed with a *terra rossa* soil, a trait that is fairly common in Levantine pottery technology. Characteristic inclusions comprise a high percentage of microfossils, including *Amphiroa*, and mollusc shell fragments (Figure 4.8). Chert is also common, together with micritic limestone with common quartz inclusions. In several cases serpentinized minerals accompany these constituents. This fabric, with a combination of the bioclasts, chalk and chert, is unique amongst the Canaanite Jar sample examined in the present study.

In terms of provenance, it was evident that, firstly, the high content of carbonates rather than quartz was characteristic of the region to the north of Israel and that, secondly, the presence of microfossils such as *Amphiroa* indicated

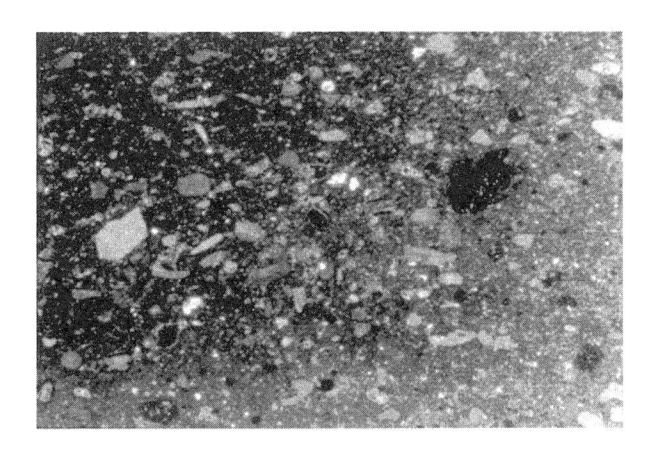

Figure 4.8 Example of Group 4 fabric. Specimen 42: with high proportion of limestone and presence of microfossils. (XPL: field width c. 10.7 mm).

that this was another Group with a coastal origin. Thirdly, the chert includes opaque minerals, which often occur as a replacement after dolomite, indicating a source in a Cenomanian chert region. On this basis, it was suggested that the source may have lain between coastal northern Israel, and Syria. However, it had not proved possible to proceed much further in circumscribing the provenance area more closely at the time of the Social Context of Technological Change Workshop in 2000.

RESULTS OF STUDY 2000–2002

We now turn to describing the current, stage of the provenance study, based on our findings made during the last two years, regarding fabric groupings and the attributions to source area.

Group 1
The location of this group on the basis of comparison with local geology remains unchanged.

Group 2
As we saw, initially we could only attribute this Group to the Coastal Plain of Israel, approximately between Ashdod and Akko. Even in the current stage of the project, it has not been possible to circumscribe very closely the southern extent of the source area. At present, we can say that Hamra-made pottery where quartz is dominant is distributed in Israel in sites located mainly along the central Coastal Plain; the southernmost known site being Yavneh Yam (Singer-Avitz and Levy, 1992). However there is a group of sites from which fabrics of this type have been recorded as dominant; comprising Aphek, Jaffa, (Kaplan 1972) and Tel Hefer, and other sites in the Coastal Plain, together with Tel Shiqmona, Tel Dor, Tel Nami, (Stern, Lewinson-Gilboa and Aviram, 1993, 358, 612, 1095–1098, 1373–

64   *Laurence Smith et al.*

1374), whose distribution extends up along the Carmel stretch of the Coastal Plain. On this basis, we can suggest that the most likely source area is the area of the Carmel coast, but possibly extending further south to about Yavneh Yam.

Group 3

This was one of the problematic groups, despite being so distinctive (Figure 4.9). We initially thought that the closest match was with the material from the central Jezreel Valley into eastern Galilee, particularly from sites to the south of Affula, which had a similarly high proportion of basaltic inclusions. But, one point was worrying: the basalt inclusions in the Affula material and in the Canaanite Amphora Group 3, appeared different (Figure 4.10). The Group 3 basalt is typically hypocrystalline, containing olivine and augite with elongated plagioclase, enclosed by a glassy phase. The homogeneity of this basalt type throughout the samples indicates a single flow of this specific type, rather than various basalt flows of different types. We realised that these characteristic should, in fact, eliminate the Jezreel Valley and the Central Jordan Valley as possible sources for

Figure 4.9 Group 3 Fabric showing hypocrystalline basalt, comprising olivine and augite columnar plagioclase, enclosed by a glassy phase. (PPL: field width c. 4.2 mm).

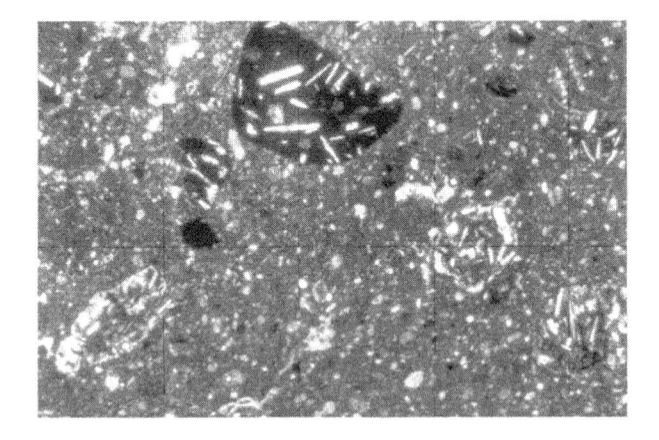

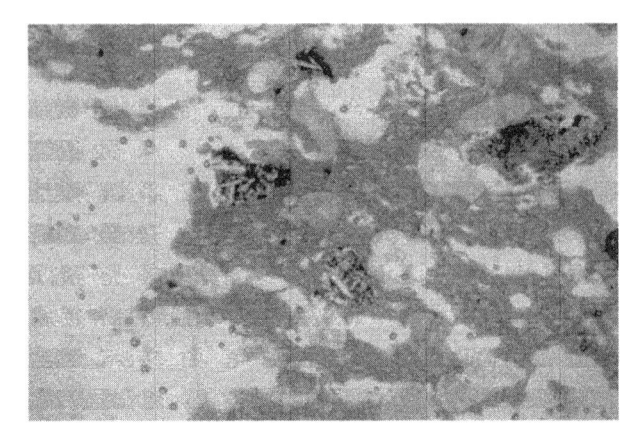

Figure 4.10 Specimen of pottery from Afule, showing typically holocrystalline basalt (PPL).

this group, as both the local drainage systems draw in basalts of various ages and different types. Hypocrystalline basalt is not recorded from these areas, where the Miocene to Pleistocene basalts are typically holocrystalline.

Whilst it is true that the general geology reflected in the inclusion assemblage of this group is typical of volcanics of Galilee or the Golan area, similar exposures appear considerably further north (Figure 4.3) in the region of the Nahr el Kebir and its opening onto the Akkar Plain, where Pliocene argils and marl, Senonian chalk and marl, and calcareous formations are found (Ponikarov 1966, sheets I–36–XXIV; I-37–XIX; Sanlaville 1977, 25, 243–280, Carte No. 1). Another indication of the source of this group has proved to be the paleontological data. The types of foraminifers within the matrix suggest that it was formed from a Neogene marl.

Fortunately from the provenance point of view, these marls do not appear in the Levant south of the Lebanese coast. Even in Lebanon, outcrops of such marls in geographically close association with volcanic outcrops appear together *only* in the 'Akkar Plain, including the region immediately north of the Nahr el-Kebir channel.

This provenance was supported by the circumstance that the only significant site known in this area, Tell Arqa, has been identified as the source of one of the Amarna letters, sent by the elders of the city of Irqata (i.e. Tell Arqa). Petrographic examination of this letter (Goren, Finkelstein and Na'aman 2003, 6–7; Goren, Finkelstein and Na'aman, in press) shows that it is of the same fabric, observed petrographically, as that of Group 3. In this way, Group 3 proved to be one of the most peripatetic of the Groups, but is now firmly attributed to the Akkar plain.

Group 4

In the initial study of the samples forming Group 4, we were able to divide the Group into the following four subgroups:

4.1.1 High percentage of ophiolite: High percentage of serpentine
4.1.2 High percentage of ophiolite: High percentage of radiolarian chert
4.2.1 High percentage of ophiolite: High percentage of replacement chert: Low percentage of radiolarian chert
4.2.2 Low percentage of ophiolite: Low percentage of replacement chert: High percentage of metamorphic rock fragments.

The general region of provenance for the Group as constituted by the first three subgroups has remained essentially unchanged. The range of inclusions is characteristic of an ophiolite complex, the latter being considered to represent oceanic crust, which has been thrust onto continental crust. Although ophiolites have a relatively extensive distribution in the eastern Mediterranean region (Whitechurch, Juteau and Montigny 1984, 306–307), the presence of reddish-stained radiolarian chert fragments in moderate percentages is a feature of the ophiolites of northwest Syria and the Hatay (c.f. Dubertret 1955, 91–92, Pl. XVI, Fig. 2). On this basis, the origin of Group 4 is most likely to be in the area stretching from the Syrian coast north of Latheqieh to the Iskenderun Bay, in Turkey.

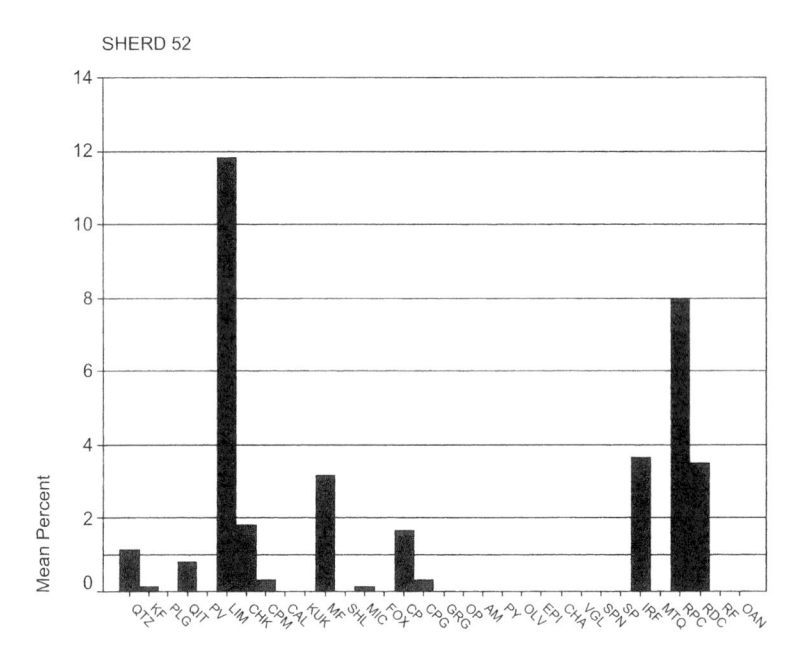

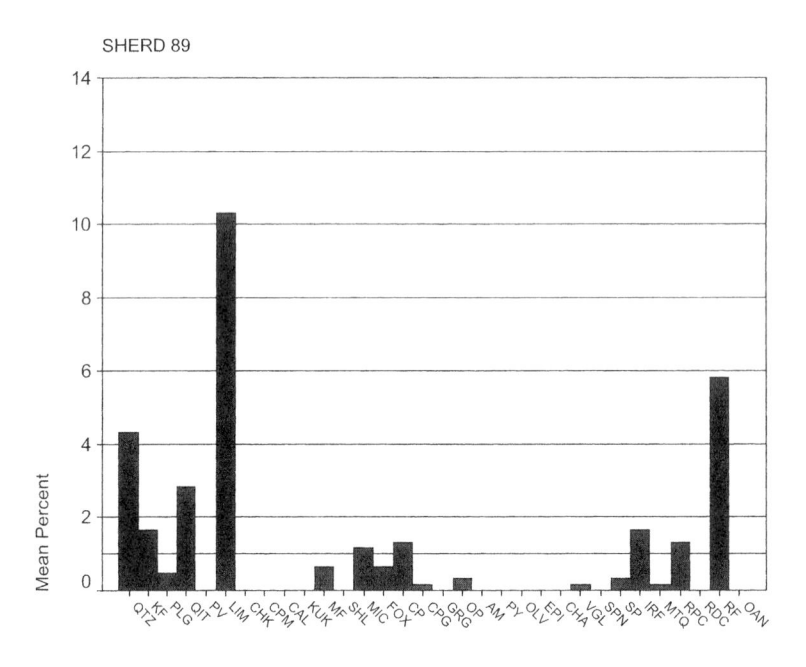

Figure 4.11 Group 4: comparison of representative sherd of subgroup 4.1.2 (top), with subgroup 4.2.2 (bottom) subsequently redesignated Group 6.

It was the final subgroup (4.2.2), that we initially began to query, since it seemed visually and mineralogically rather different from the previous three, particularly in having (Figure 4.11) a lower abundance of the minerals and rock fragments characteristic of the ophiolite complex, and in particular, very low content of the radiolarian chert, which we had come to regard as one of the main characteristics of the Group. We subsequently analysed more samples of the fabric falling into this last subgroup, this having already been put forward as a separate fabric at Amarna by Margaret Serpico. When the 'subgroup' was enlarged with further specimens, it did start to appear consistently different from the other subgroups in the low percentage, or more often absence, of the radiolarian chert, (although replacement chert was still present); in the frequent high proportion of metamorphic quartz grains, and in the presence of altered basaltic inclusions, such as spillites, and the presence of probable fragments of plagiogranite. It was, therefore, decided to remove 4.2.2 from Group 4, and designate it as a new fabric group: Group 6. We then turned to the chemical analysis, firstly to see whether this would support the separation of 4.2.2 as a new Group.

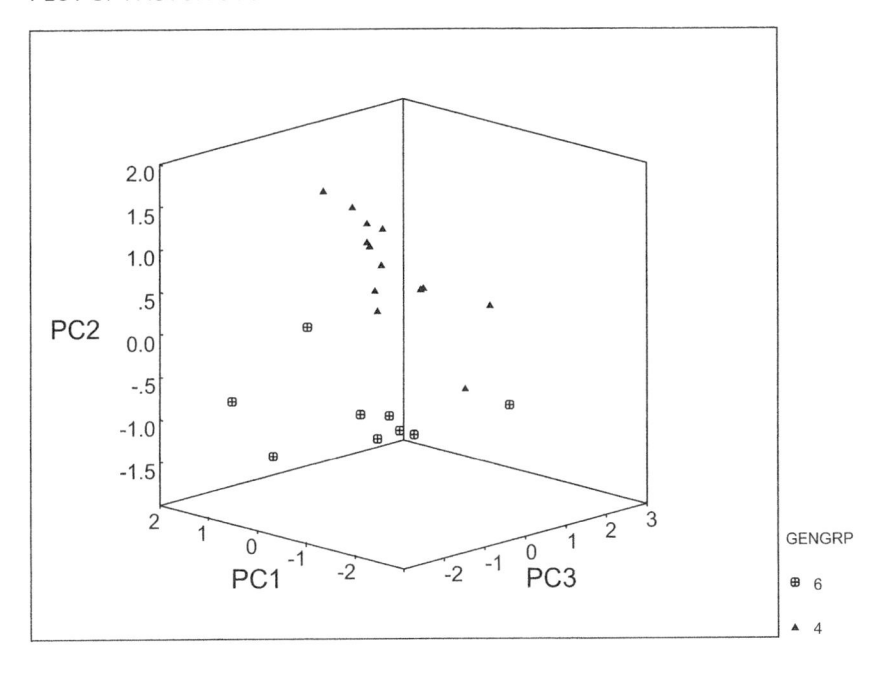

Figure 4.12 Plot from Principal Component Analysis of chemical data showing relationship between Group 4 and Group 6 fabrics.

When the chemical data was examined with this in mind, it showed that, when ICP-AES data alone was considered under principal components analysis (PCA), Fabric Group 6 appeared quite distinct from Group 4 (Figure 4.12). The splitting of Fabric Group 6 from Fabric Group 4 was supported by the subsequent runs using Cluster Analysis on the joint ICP-AES/NAA data, since Fabric Group 6 clustered consistently separately from Fabric Group 4, although there was a slight overlap with Fabric Groups 2 and 3. Despite these similarities with some of the other Fabric Groups, the chemical analysis, for the most part, supported the separation of Fabric Group 6 from Fabric Group 4 as originally constituted.

Group 6

Given the decision to separate out Group 6, the question of its geographical origin then arose. If, based on the initially observed similarity of Group 6 to Group 4, one were to assume that the vessels in Group 6 came from the same area as those classed as Group 4, there was not a close match with the Ras Shamra

Figure 4.13 Example of Group 6 fabric. Specimen 89: showing abundant quartz and feldspars, with some grey clay pellets, but lacking main characteristics of Group 4. (XPL: field width c. 10.7 mm).

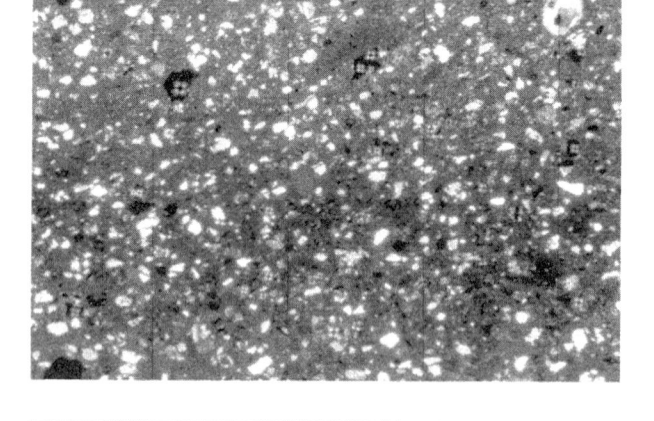

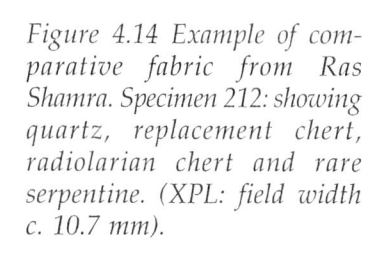

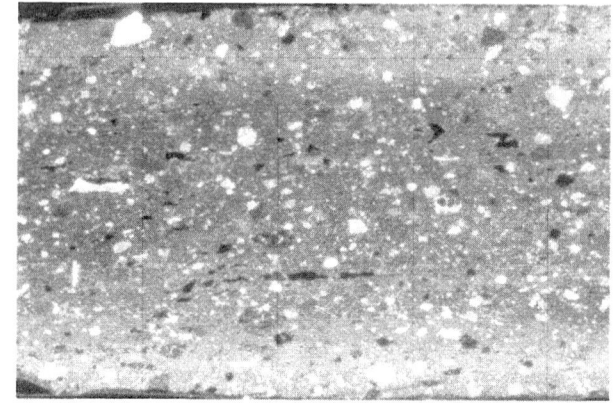

Figure 4.14 Example of comparative fabric from Ras Shamra. Specimen 212: showing quartz, replacement chert, radiolarian chert and rare serpentine. (XPL: field width c. 10.7 mm).

samples (Figures 4.13 and 4.14). Even if other areas a little further north, such as the Amuq, were considered also as potential source areas, the radiolarian chert was still more evident in sections from sherds considered to be of local pottery than in our Group 6. Also, spillites and plagiogranite are extremely uncommon in the north Syrian and Hatay ophiolites, which are mainly of ultrabasic composition. Examples of local pottery from further northwest into Cilicia were predominantly composed of serpentine and peridotitic fragments. So, it seemed that an origin in the Northwest Syrian-Cilician region for this subgroup was unlikely.

However, it was realised that the features of the subgroup were more characteristic of the Cypriote ophiolite of the Troodos and Mamonia complexes in having replacement, but virtually no radiolarian chert. Although radiolarian *shales* occur to a limited extent in the Mamonia area (Gass *et al.* 1994; Geological Survey Department, Cyprus 1995), most of the radiolarian *chert* has been removed from these Cypriote ophiolites by erosion. They do, however, have a relative abundance of spillites.

To explore this possible Cypriote origin, examples of local pottery were obtained from several sites on Cyprus. Samples with the closest similarity to Group 6 were collected for comparison. Sherds from Kalavasos-Ayios Dhimitrios (South 1997), and Sandiha, in particular, were considered on the basis of low power examination to resemble the Group 6 specimens, although they had seemed to differ from the other subgroups in Group 4 as originally constituted. When compared in thin section, unfortunately, the degree of matching was not great; the Cyprus local samples available to us tended to be rather coarser grained, and to have a higher proportion of weathered and altered basaltic inclusions (spillites). However, the basic characteristics of the Group 6 did occur, including the presence of volcanic glass, serpentine, igneous rock fragments, and the virtual absence of radiolarian chert (Figure 4.15).

Regarding the attribution to Cyprus, the results of the chemical analysis were not absolutely clear, but showed that there was clearly some overlap in composition between Fabric Group 6 and the Cyprus local sample. In PCA, two Cyprus specimens fell near to Group 6. However, nearly all specimens considered most likely to have been made on Cyprus clustered separately from Fabric Group 6. The Discriminant Analysis of the relocated clusterings supported the distinctiveness of most of the Cyprus sample. In the Discriminant plot (Figure 4.16) the Cyprus local cluster (Cluster 5) appeared separate on Discriminant Function 1 and Discriminant Function 3 from the clusters containing Fabric Group 6.

This result complemented that of the thin section study, in showing that although some of the Cyprus local comparatives available to us were compositionally close to Fabric Group 6, most were distinct from the latter. Similarly, the thin section data showed that the Cyprus local specimens were geologically related in that they were from an ophiolite complex essentially lacking radiolarian chert, but the specific constituents and texture of the fabric were not identical to that of Fabric Group 6. This may indicate that the sites from which samples were

SHERD 278 (Kalavasos Ayios Dhimitrios)

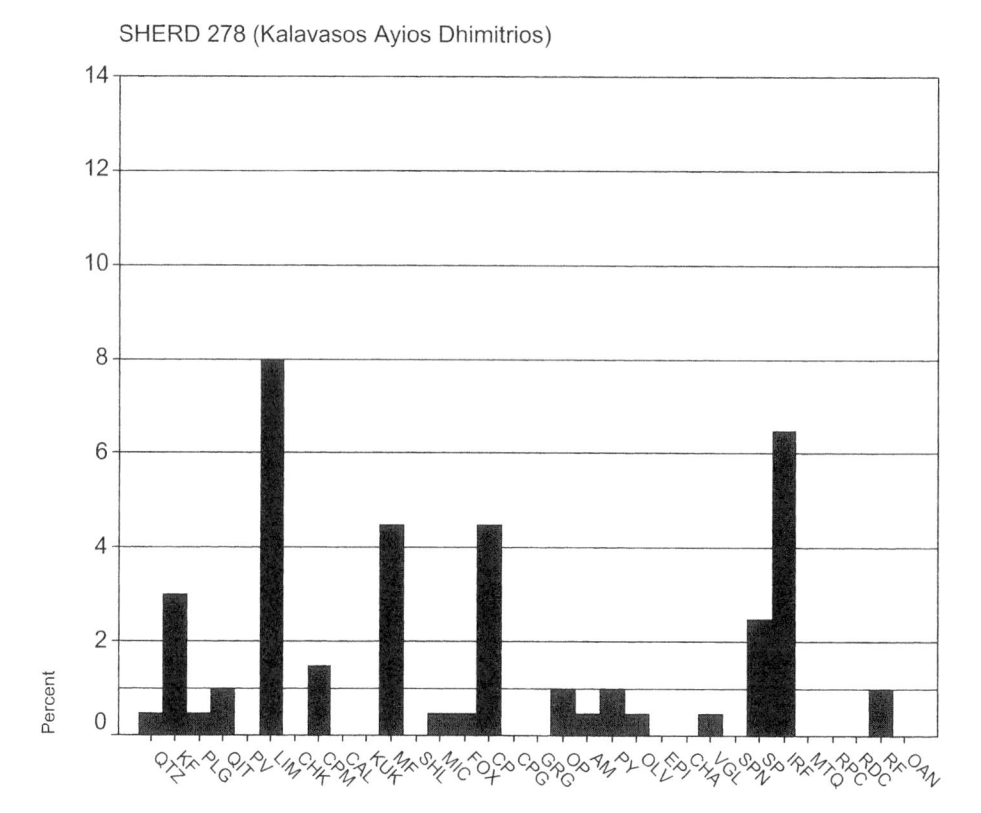

Figure 4.15 Barchart showing constituents of specimen from Cyprus: Sherd 278, Kalavasos Ayios Dhimitrios.

taken were too far inland towards the Troodos area, whereas more suitable comparative material would be from sites representative of the geology nearer the coast, such as Amatos or Hala Sultan Tekke.

Samples from Hala Sultan Tekke were included in the main thin section study, but these were found to be imports, and so were not representative of local Cypriote fabrics. A preliminary comparison between thin sections of Group 6 and of sherds from Enkomi (from the IATU thin section collection) was carried out on the basis of visual inspection only. Some similarities were noted between Group 6 and the Enkomi samples, but no exact matches in both constituents and fabric texture were found. Given the scope of the present study, it has not been possible at the time of writing to undertake more detailed comparisons, based on point-counting and chemical analysis, of Group 6 with a larger sample of local ceramics from Cypriote coastal sites.

Plot of Discriminant Functions

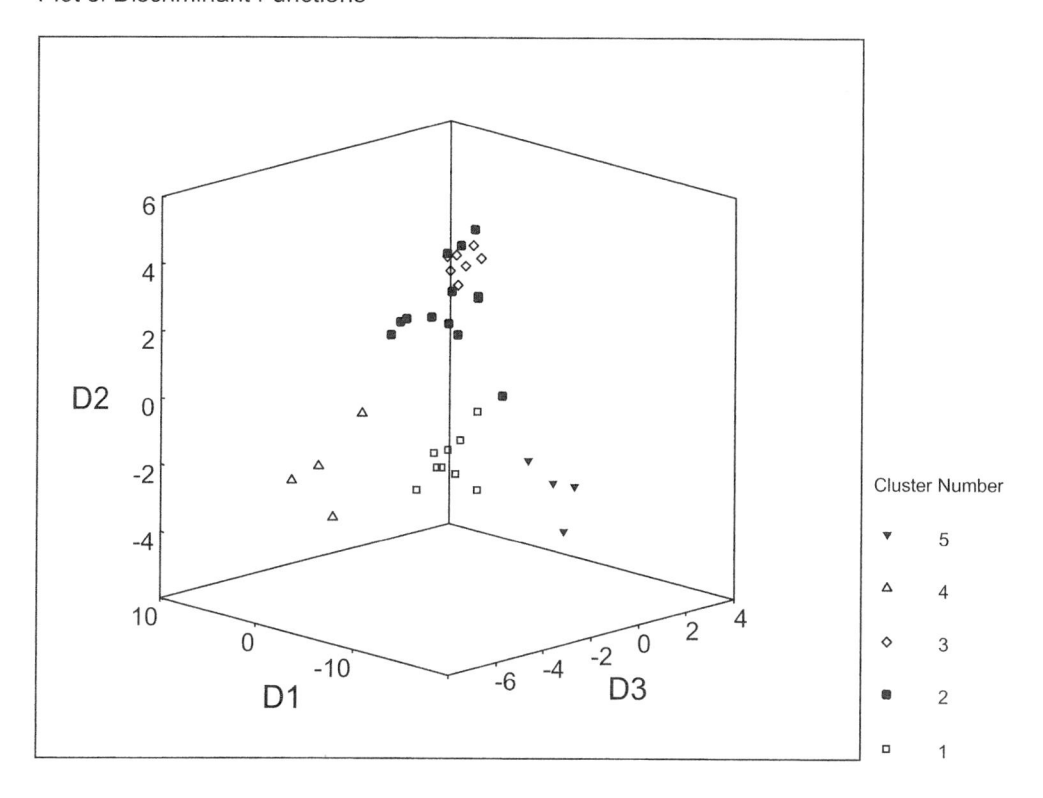

Figure 4.16. Plot from Discriminant Analysis of groupings from Cluster Analysis of chemical data showing relationship between Cyprus local specimens (Cluster 5) and Group 6 fabric specimens (Clusters 1 and 4).

Group 5

With regard to the remaining Group to be discussed, Group 5, we were able to suggest that the source for this Group may have lain generally in coastal Lebanon or Syria. Since we knew that the source of the material should be in an area where exposures of limestone, chalk and chert appear together with Pleistocene to recent beach deposits of mainly calcareous character this meant that our samples should be related *a priori* to the coastal area near Akko or to the north of this area.

Subsequent work has led us to realise that other components within the inclusion assemblage may delimit the source rather more closely. Firstly, in the Levant generally, chert and associated chalk, is related either to the Senonian or the Eocene exposures. Such rock types occur predominantly between Tyre and Sidon, and again north of Tripoli although small exposures exist also as far south as the area to the east of Akko. However, this limits the possible sources for

72 *Laurence Smith et al.*

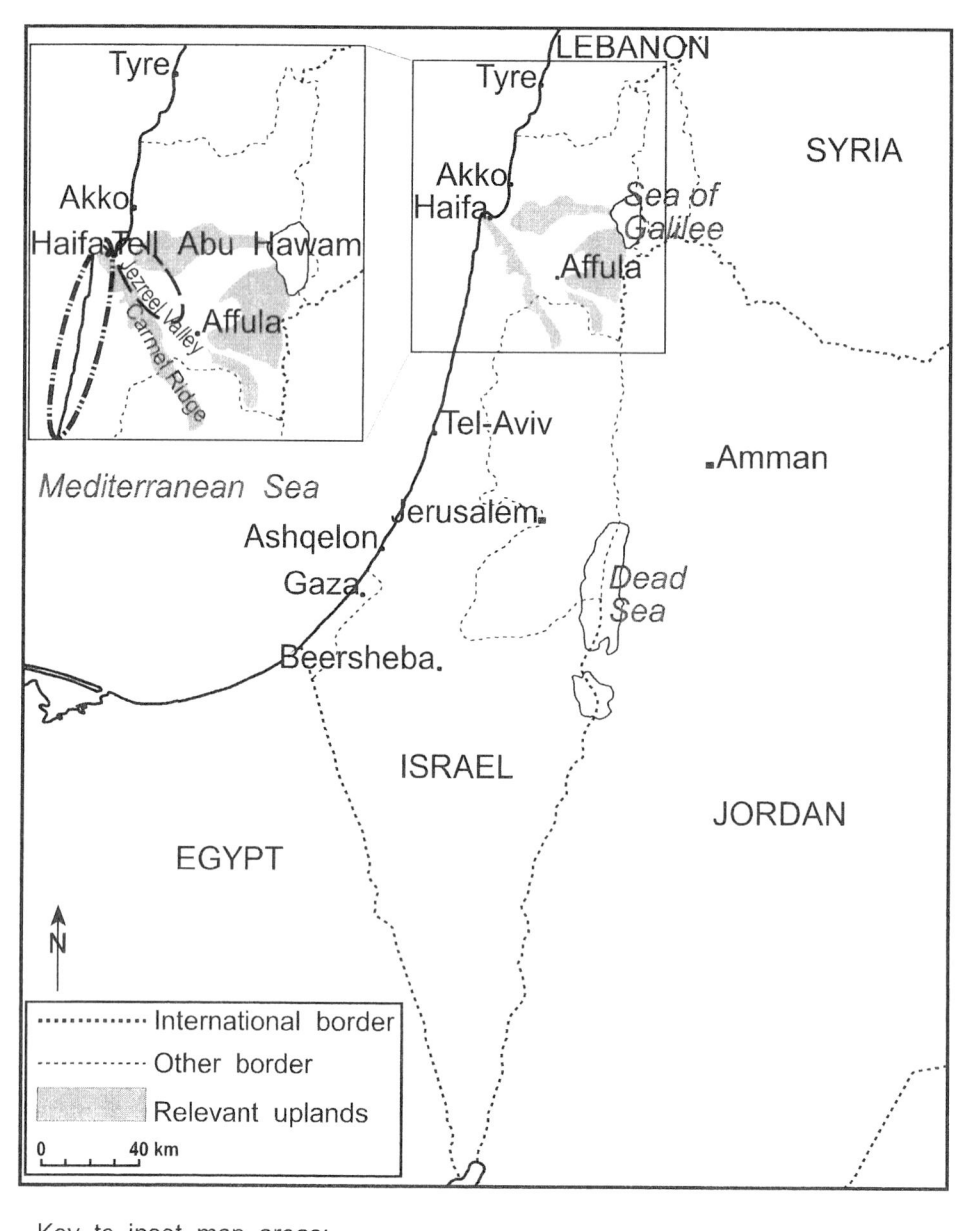

Figure 4.17 Map showing provenance areas for Groups 1 and 2 on basis of geology. (Map drawn by L. Smith using 'Professional Draw' package, by Gold Disk Inc.).

Group 5 to these regions alone. Secondly, in Lebanon, the beach sand dunes near Tyre still contain a minor component of quartz, but the sand is made essentially of carbonates, mostly from bioclasts. Further north the sand dwindles and in Sidon it is virtually absent. Further north still, it appears again in patches and becomes abundant again at the Akkar Plain, in the Lebanon/Syria border region (Sanlaville 1977, 161–164).

On this basis, we still assign the entire Group 5 to a locality or localities in the general area of the Lebanese coastal plain. However, there is some variability within Group 5 in the percentages of quartz, limestone fragments and microfossils. We still need to do some final checking on this point, but at present, the area between Akko and Sidon seems to be the best candidate for the origin of part of this group. Because of the change from quartz-dominated to carbonate-dominated sands as one goes northwards up the coast (c.f. Nir 1989), it can be suggested that in samples where quartz occurs at lower frequencies together with the highest proportion of limestone and microfossils (in Subgroup 5.2.1 and 5.2.2), the more northern Lebanese/Syrian coast, extending from about Sidon to the Akkar plain, can be regarded as a possible origin as well.

CONCLUSIONS

Now it is possible to offer the following summary of the provenances of the Amarna and Memphis Canaanite Jar sample included in the present Project (see also Figures 4.17 and 4.18):

Group 1: The seaward end the Jezreel Valley to the north-east of the Carmel Ridge, west of Affule

Group 2: Coastal plain, in Carmel region

Group 3: Present-day Lebanon-Syria border area, specifically inland 'Akkar Plain region

Group 4: Generally from Syrian coast north of Latheqieh to Iskenderun Bay, Turkey, most likely in the Baër-Bassit region

Group 5: Generally Lebanese coastal plain. Higher-quartz sub-group: area between Akko and Sidon. Low-quartz subgroups: Lebanon/Syria region, coastal 'Akkar Plain

Group 6: Southern coast of Cyprus, generally between Paphos and Enkomi.

Thus, our refinement of provenance has enabled us to narrow the source areas of two of the Groups. Group 2 changed from a broad distribution that could have included the central coastal and central inland portions of present-day Israel to more specifically the northern part of the coast, and the Group 5 region was narrowed to the coastal Lebanon/Syria border. This, taken in conjunction with the migration of Group 3 vessels from the Jezreel Valley/eastern Galilee region to the inland Akkar Plain, again in the Lebanon/Syria border region, shifts the overall distribution of our Canaanite amphorae sample during this period considerably further north, towards the northern half of the Levant. We now see

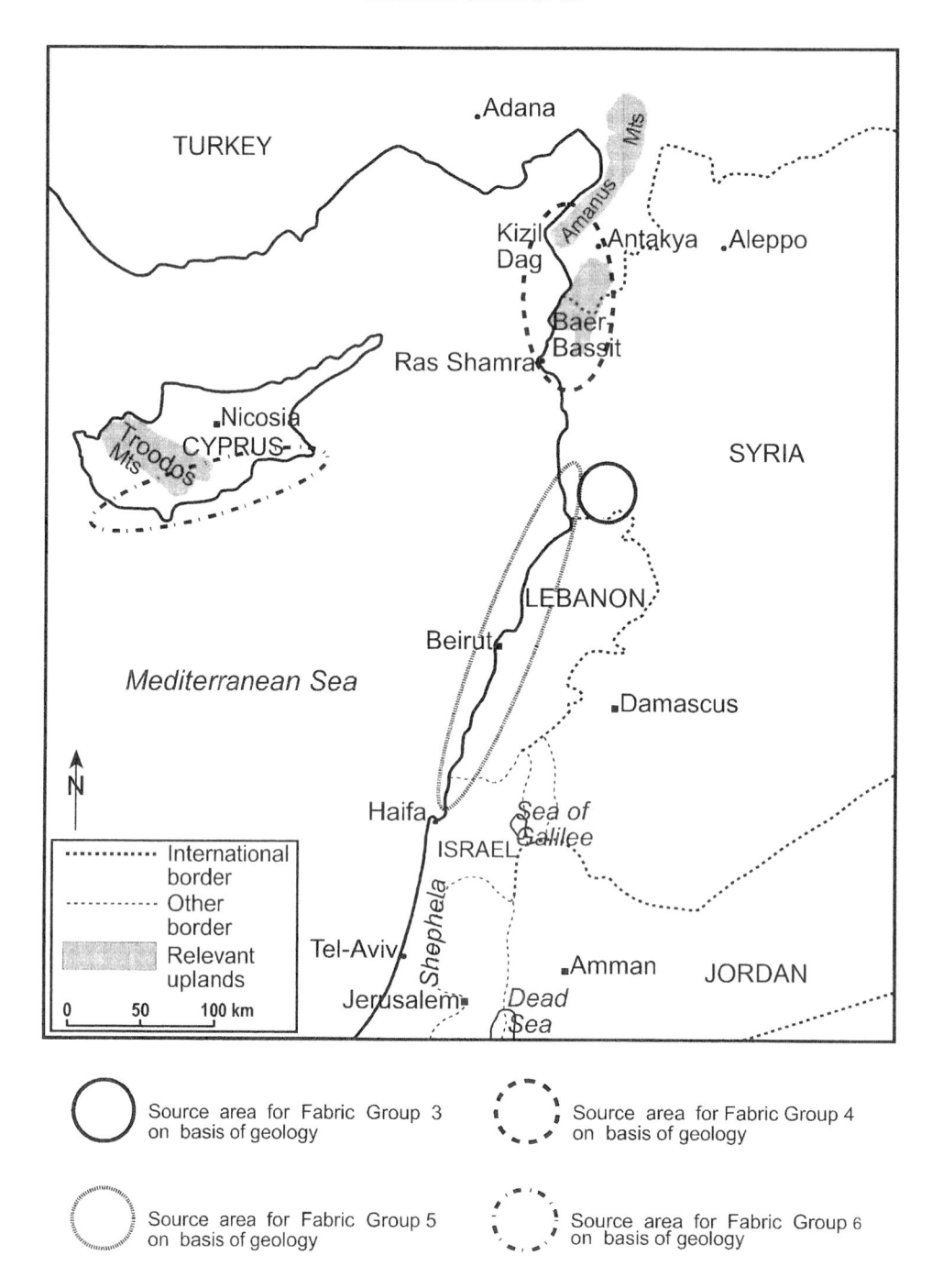

Figure 4.18 Map showing provenance areas for Groups 3, 4, 5 and 6 on basis of geology. (Map drawn by L. Smith using 'Professional Draw' package, by Gold Disk Inc.).

that Group 1, originally in the middle of the potential distribution of the source areas, now appears to represent almost the southernmost region of origin.

ACKNOWLEDGEMENTS

The Egyptian data derive from sherds from Memphis and Amarna exported by permission of the Supreme Council for Antiquities, Egypt. These were supplemented by sherds from earlier excavations at Amarna, which are now in the Petrie Museum of Egyptian Archaeology. Some 40 comparative samples were collected in Israel by permission of the relevant excavators and the Israel Antiquities Authority; 6 samples from Ras Shamra were collected by permission of the Department of Near Eastern Antiquities, the Louvre, Paris; and 10 samples were collected in Cyprus, by permission of the Director of Antiquities and Ian Todd and Alison South, the excavators of Kalavassos. We should like to thank all those involved in granting these permissions, for without them the project would not have been possible. Funding was provided by the Egypt Exploration Society, McDonald Institute for Archaeological Research, Cambridge, Wainwright Fund of the Oriental Institute in the University of Oxford, British Academy and Society of Antiquaries of London. Collaborators on residue analysis are C. Heron and B. Stern of the University of Bradford, supported by a NERC grant; and on the petrological study, Judith Bunbury of Department of Earth Sciences, University of Cambridge. Thanks are due to the following members of staff of the Department of Archaeology for assistance during the project: Dr. C. French, Dr. C. Shell, Ms. J. Rippengal, Ms. J. Miller. Thin sections were prepared by Ms. K. Knowles and Mr. R. Winterbottom of the Department of Archaeology, University of Southampton, and by Mr. S. Houlding, Department of Earth Sciences, UCL.

NOTES

[1] The first author was responsible for the fabric descriptions and obtaining quantitative data by point-counting, with help from Dr.J.Bunbury, Department of Earth Sciences, University of Cambridge, in the identification of certain mineral and rock fragments.
[2] The analysis was run at Royal Holloway College, London under the direction of Dr. J.N.Walsh.

REFERENCES

Al-Dayel, O. A. F., 1995, Characterisation of ancient Egyptian ceramics by Neutron Activation Analysis. Unpublished Ph.D. dissertation, University of Manchester.
Beydoun, Z.R., 1977, The Levantine countries: the geology of Syria and Lebanon (maritime regions), in A. E. M. Nairn, W. H. Kanes, and F. G. Stehli (eds), *The ocean basins and margins. 4A: The eastern Mediterranean*, 319–353. Plenum Press, New York and London.
Bourriau, J.D., 1990, Canaanite Jars from New Kingdom Deposits at Memphis, Kom Rabi'a. *Eretz-Israel*, 21, 18–26

Bourriau, J.D. and Nicholson, P.T., 1992, Marl Clay Pottery Fabrics of the New Kingdom from Memphis, Saqqara and Amarna, *Journal of Egyptian Archaeology* 78, 29–91.

Bourriau, J. D., Smith, L. M. V. and Serpico, M., 2001, The provenance of Canaanite Amphorae found at Memphis and Amarna in the New Kingdom, in A. Shortland (ed.), *The social context of technological change: Egypt and the Near East 1650–1550 BC*, 113–146. Oxbow Books, Oxford.

Dubertret, L., 1955, Géologie des roches vertes du nord-ouest de la Syrie et du Hatay (Turkie). *Notes et Mémoires sur le Moyen-Orient* VI, 1–224. Paris, Muséum National d'Histoire Naturelle.

Gass, I. G., MacLeod, C. J., Murton, B. J., Panayotou, A., Simonian, K. O. and Xenophontos, C., 1994, *The Geology of the Southern Troodos Transform Fault Zone*. Cyprus Geological Survey Department (Memoir No. 9), Nicosia.

Geological Survey Department, Cyprus, 1995, *Geological map of Cyprus.* 1:250,000. Geological Survey Department, Ministry of Agriculture, Natural Resources and Environment, Government of Cyprus.

Getzov, N., 1993, Hurvat 'Utza, *Excavations and Surveys in Israel*, 13, 19–21.

Goren, Y., Finkelstein, I. and Na'aman, N., 2003, The expansion of the Kingdom of Amurru according to the petrographic investigation of the Amarna Tablets, *Bulletin of the American School of Oriental Research*, 329, 1–11.

Goren, Y., Finkelstein, I. and Na'aman, N., (in press), *Inscribed in clay I, provenance study of the Amarna letters and other ancient Near Eastern texts*. Institute of Archaeology, Tel-Aviv University, Tel-Aviv.

Goren, Y., Bunimovitz, S., Finkelstein, I. and Na'aman, N., 2003, The location of Alashiya: new evidence from petrographic investigation of Alashiyan Tablets from El-Amarna and Ugarit, *American Journal of Archaeology*, 107, 233–255.

I.F.P.-C.N.E.X.O., 1974, *Carte géologique et structurale des bassins Tertiaires du domaine Méditerranéen*. 1st edition 1:250,000. L'Institut français du Pétrole, Le Centre National pour l'Exploitation des Océans, L'Institut National d'Astronomie et de Géophysique. Éditions Technip, Paris.

Kaplan, J., 1972, The archaeology and history of Tel Aviv-Jaffa, *The Biblical Archaeologist*, 35, 66–95.

Nicholson, P.T., and Rose, P., 1985, Pottery fabrics and Ware Groups at el-Amarna, in B. Kemp (ed.), *Amarna Reports II*, 133–174. Egypt Exploration Society, London.

Nir, Y.,1989, *Sedimentological Aspects of the Israel and Sinai Mediterranean Coasts*. Geological Survey of Israel internal report, Jerusalem (in Hebrew).

Ponikarov, V.P., (ed.), 1966, *The Geological Map of Syria, 1:200,000 (19 sheets with explenatory notes)*. Ministry of Industry, Syrian Arab Republic, Damascus.

Sanlaville, P., 1977, *Étude Geomorphologique de la Region Littorale du Liban*. Publications de l'Universite Libanaise, Beirouth.

Serpico, M., 1996, *Mediterranean resins in New Kingdom Egypt: a multidisciplinary approach to trade and usage*. Unpublished Ph.D. University College, London.

Serpico, M. and White, R., 2000, The botanical identity and transport of incense during the Egyptian New Kingdom, *Antiquity* 74, 884–97.

Serpico, M., Bourriau, J., Smith, L., Goren, Y., Stern, B., and Heron, C., 2003, Commodities and Containers: A Project to Study Canaanite Amphorae Imported into Egypt during the New Kingdom, in M. Bietak (ed.), *The Synchronisation of Civilisations in the Eastern Mediterranean in the Second Millennium B.C. II. Proceedings of the SCIEM 2000 – EuroConference, haindorf, 2nd of May – 7th of May 2001*, Österreichische Akademie der Wissenschaften,Wein. 365-76.

Singer-Avitz, L. and Levy, Y., 1992, An MBIIa kiln at the Naḥal Soreq Site. *Atiqot*, 21, 9–14. (in Hebrew).

Sivan, D., 1996, *Paleogeography of the Galilee coastal plain during the Quaternary*. Report GSI/18/96. Geological Survey of Israel, Jerusalem.

Smith, L. M. V., Bourriau, J. D. and Serpico M., 2000, The provenance of Late Bronze Age transport amphorae found in Egypt, *Internet Archaeology* 9 (http://intarch.ac.uk/journal/issue9).

South, A. K., 1997, Kalavasos-Ayios Dhimitrios 1992-1996. *Report of the Department of Antiquities, Cyprus*, 151–175.

Stern, B., Heron, C., Serpico, M. and Bourriau, J., 2000, A Comparison of Methods for Establishing Fatty Acid Concentration Gradients across Potsherds: A Case Study using Late Bronze Age Canaanite Amphorae, *Archaeometry* 42, 399–414.

Stern, B., Heron, C., Corr, L., Serpico, M. and Bourriau, J., 2003, Compositional Variations in Aged and Heated Pistacia Resin found in Late Bronze Age Canaanite Amphorae and Bowls from Amarna, Egypt, *Archaeometry* 45, 457–69.

Stern, E., Lewinson-Gilboa, A. and Aviram J. (eds.), 1993, *The New Encyclopedia of Archaeological Excavations in the Holy Land*, 4 Vols. Simon and Schuster, New York and London.

Whitechurch, H., Juteau, T. and Montigny, R., 1984, Role of Eastern Mediterranean ophiolites (Turkey, Syria, Cyprus) in the history of the Neo-Tethys, in J. E. Dixon and A. H. F. Robertson (ed.) *The Geological Evolution of the Eastern Mediterranean*, 301–317. The Geological Society, Oxford.

Chapter 5

The Beginnings of Amphora Production in Egypt

Janine Bourriau

Abstract
There is abundant evidence, both archaeological and textual, of the large scale and specialised production of amphorae during the New Kingdom. However, the evidence showing that amphorae of Egyptian manufacture were actually an innovation of c.1550 B.C. is only now apparent. What factors can we see behind this change? What are the implications for the organisation of pottery making in the New Kingdom? What is the relation-ship between the morphology of Egyptian amphorae and Canaanite Jars of the Middle Bronze and Late Bronze Age? Can we, by analysing the raw material of the amphorae, suggest the region where they were first made? These are some of the questions which will be addressed in this paper, using unpublished data from the excavations of the Egypt Exploration Society at Memphis, Kom Rabi'a.

There is abundant evidence, both archaeological and textual, of large scale and specialised production of amphorae in Egypt during the New Kingdom. Detailed studies have been written (Hope 1989; McGovern 1997; Bourriau, Smith and Nicholson 2000) and the evidence has frequently been rehearsed elsewhere, usually in discussions of wine production. Wine was the commodity most frequently carried in the amphorae, as indicated by their hieratic labels. These are ink inscriptions written on the amphorae after they were filled, and which have survived in greater numbers than the vessels themselves (Lesko 1977; Meeks 1993; McGovern, Fleming and Katz 1996; Murray 2000). Not every amphora was labelled, or has a label which survives and can be read, but at their most complete the labels can tell us the commodity; its source and destination; the date (by regnal year) when it was sealed; its dedicator, if any; for wine, the name of the vintner; and the quantity of the commodity in the vessel (Koenig 1979; Koenig 1980; Tallet 1998).

Recent advances in the study of this material have generated fresh ideas about economic and social organisation, as well as answering those conventional questions of origin and date which are still the foundation upon which all other hypotheses rest.

The most important advance has been the integration of the study of the vessels themselves, their morphology, material and technology, with that of the inscribed labels, seals and sealings. One of the first studies to do this was Colin Hope's discussion of the amphorae from Malkata (Hope 1978). The data of the linguist and the ceramicist complement each other: thus, the identification of a place name may be made easier if the vessel is known to be made of clay from the Western Oases, or re-use of a vessel, not attested by its inscription, may be established if its shape or fabric is known to be linked, in its primary use, to a particular commodity. Until some 30 years ago, by which time most amphora labels in museums had already been collected, it was common practice for excavators to record them without reference to the amphorae upon which they occurred. The labels would be removed, often at the expense of the vessels themselves, as the fresh breaks testify. Similar treatment was once accorded to the stamped handles of Greek transport amphorae of a later period. Happily, in both Egypt and the Aegean this approach has been largely abandoned as far as field recording is concerned. However, there is always a danger that information may be lost if a holistic approach is not taken and the label goes off to the epigrapher, becoming transformed by an archaeological miracle into a small find, while the rest of the amphora stays with the ceramicist. Regrettably, publications do still appear in which each inscription is treated as if it were an ostracon, and the sherd merely the writing surface, rather than all that remains of the object to which the text refers. It must be said in fairness that if the collection of labels is an old one, nothing may remain but inscribed body sherds and in this case even if the fabric can be identified, this alone cannot fully recover the information which has been lost. Nevertheless, current research on previously-excavated material from Deir el-Medineh in the French Archaeological Institute in Cairo (Bavay, Marchand and Tallet 2000) and from Malqata in the Metropolitan Museum of Art (McGovern 1997), together with study of newly-excavated pottery from Thebes (Guksch 1995; Hope 2002, 102–4), is demonstrating the value of this new approach.

Advances have also been made in the study of fabrics (Hope, Blauer and Riederer 1981; Nicholson and Rose 1985; Bourriau and Nicholson 1992; Nordström and Bourriau 1993; Bourriau, Smith and Nicholson 2000); technology and shape (Hope 1989); and residues (McGovern 1997; Murray 2000).

As a result, we have new tools with which to tackle the most obvious questions: when did large-scale amphora production begin in the New Kingdom, and why did this occur? If we can answer the first question, we may be better able to answer the second.

Many Egyptian amphorae carry, not only ink inscriptions with regnal years, but also, on the handles, pre-firing stamp-impressions with the name of a King. Similar impressions were also made in the great cones of mud used to cover the jar-stopper, but they rarely survive still attached to the jar, and where they do (as in the case of Tutankhamun) they conceal the rim and neck of the vessel. The earliest pre-firing stamped handle-impression with a royal name known to me is of Tuthmosis I (1504–1492 B.C.); an example was found at Serra East in Nubia (Hughes 1963, 129; Hope 1989, 93). Although we cannot say for certain what the fabric was (Hughes described it only as cream-slipped), we can assume an Egyptian origin, and manufacture under royal control. Hope (1989) surmises that it belonged to an amphora of the type illustrated in his Fig.1.1 or 1. 2.

For earlier amphorae, we can turn to the evidence from the Egypt Exploration Society's excavations at Memphis, Kom Rabi'a. The site is a small mound providing stratified remains of domestic housing covering the period from the mid-13th Dynasty to the early Third Intermediate Period (c.1700–1070), so including a pottery sequence for the whole of the New Kingdom (Bourriau and Eriksson 1997).

Between the Second Intermediate Period level (VI) and the first New Kingdom structural level (IV) were layers of sand (Level V). These were deposited over a period of time, judging by differences in the colour, texture and contents (Giddy 1999, 3). In the deposits within Level V a change in the ceramic repertoire becomes evident, affecting just a few classes of vessel. The changes have been discussed in detail elsewhere (Bourriau 1997, 159–182; Bourriau and Eriksson 1997, 95–120) but in general terms pottery types already familiar from Upper Egypt, and specifically the Theban region in the 17th Dynasty (1640–1550 B.C.), appear here in northern Egypt for the first time and become in the next level, Level IV, the dominant ceramic style. Examples of this 'new style' appear in Level V in a repertoire otherwise familiar from other sites in the region, Lisht and Dahshur providing parallels from recent excavations. This repertoire belongs emphatically to the Middle Kingdom; it is present at Kom Rabi'a in the earliest excavated stratum (Level VIII), of the mid-13th Dynasty, evolving without a break until the sand layer is reached.

In Context 707, within the sands of Level V (Bourriau and Eriksson 1997, Fig.1), lay a large shoulder sherd of an amphora, with handle attached (Figure 5.1). The clay was unequivocally of Marl D (Nordström and Bourriau 1993, 181–

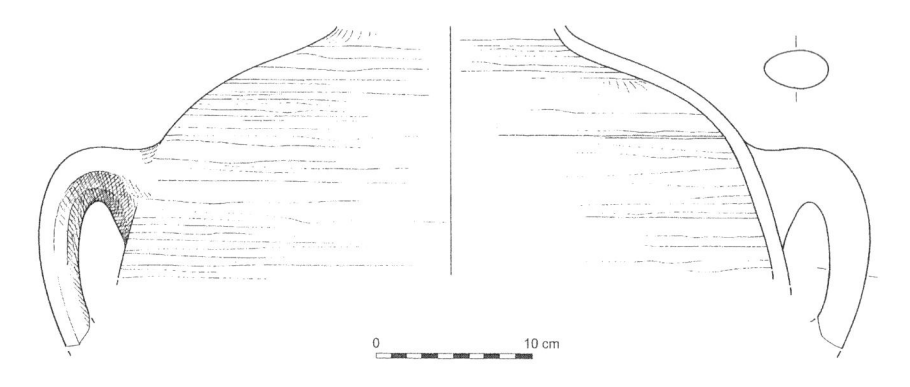

Figure 5.1 Fragment of an Egyptian amphora in Marl D from Kom Rabi'a, Context 707.

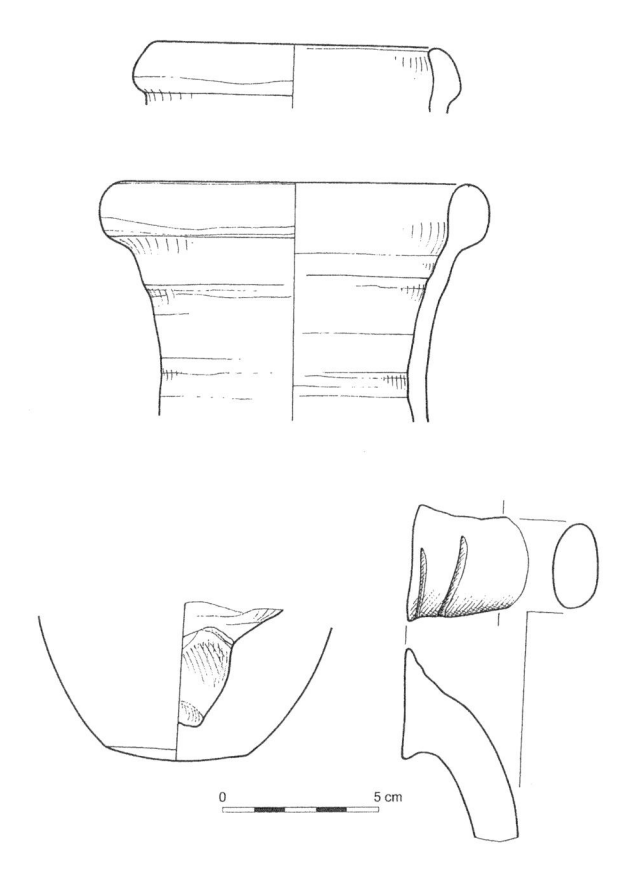

Figure 5.2 Diagnostic sherds from amphorae of Marl D, Kom Rabi'a, Level IV.

2), the most common fabric of Egyptian amphorae throughout the 18th Dynasty. The vessel was wheel thrown and the surface left uncoated. In the same deposit were further examples of the 'new' ceramic style – shallow, footed bowls, the drinking cups of the New Kingdom, so appropriately found with an amphora. A random sample consisting of 23 rim sherds from Context 707 (Bourriau 1991b) revealed that 52% were in the Middle Kingdom ceramic style and 48% in that of the New Kingdom.

From this evidence, it seems clear that we have, just before the new building phase (IV), which is orientated quite differently from the previous building phase below the sand, and whose beginning can be dated by objects and ceramics to the two earliest reigns of the 18th Dynasty, a wheel-made Egyptian amphora. Nor does it look like a prototype but rather a confidently made, familiar form. This is still true even though the exact date and duration of Level V itself are still a matter for discussion.

Contexts within Level IV produced further diagnostics: rims, handles and a base, all of Marl D and all of a single amphora type (Figure 5.2). The vessel form

82   *Janine Bourriau*

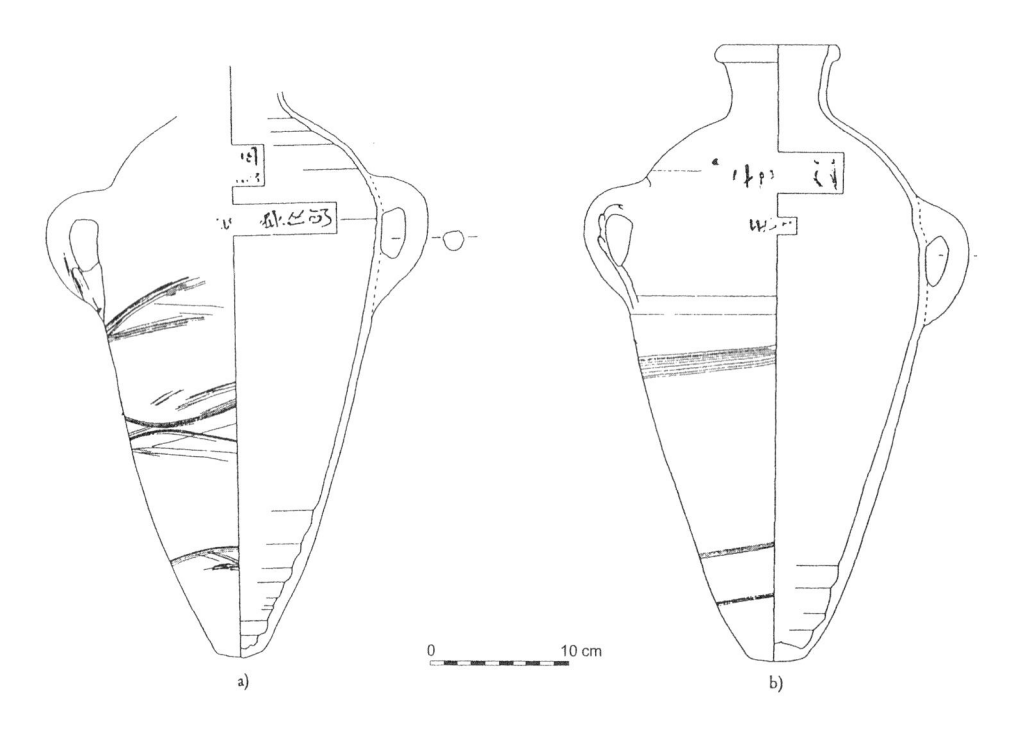

Figure 5.3 Two amphorae of Marl D from Thebes, Tomb of Nakht Min (Guksch 1995, Abb.36, a and b).

can be reconstructed with reference to the complete example published by Heike Guksch (1995) from the tomb of Nakht Min, who served Queen Hatshepsut (1473–1458 B. C.) (Figure 5.3). Also in Level IV was a 'pilgrim flask' of Marl D (Figure 5.4). This conjunction is interesting because pilgrim flasks appear in both Egypt and Israel in association with Canaanite jars and made of the same set of Canaanite fabrics. Whatever their function it was intimately connected with the amphorae. Perhaps they held a commodity routinely added to the jar's contents – a sweetening agent, for example. The total number of Marl D sherds in Level IV is small but, measured by weight, Marl D shows an increase through the stratum from 4.5% to 13.2% of the total New Kingdom ceramic.

There is no trace of either ink inscriptions or stamp impressions on this material. The absence of ink inscriptions can be explained by the damp conditions at Memphis, situated as it is beside the Nile. It is one of the major distortions of the evidence that all jar labels have been recorded from dry sites and none has survived in the damp soil of the region of Egypt where most of the wine was actually produced, i.e. the Delta. The absence of stamp-impressions, on the other hand, may be due to a combination of several factors: the small amount of material excavated; the fact that only a proportion of the amphorae were ever stamped; and the character of the contexts, in which they were found.

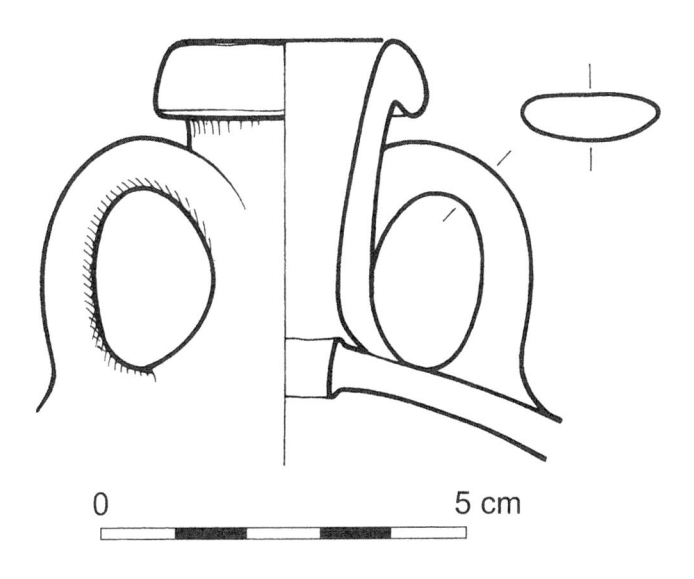

Figure 5.4 Upper part of a "pilgrim flask" of Marl D from Kom Rabi'a, Level IV.

Most stamped amphorae have been found in the 'élite' burials of officials (especially those who served the King) and favoured craftsmen such as those at Deir el-Medineh, or in the storerooms of temples and palaces. Despite its later date, the recently-excavated site most closely comparable with Kom Rab'i a is the workmen's village at Amarna. Sherds of Egyptian amphorae (Rose's Type 21) were rare there (Rose 1984) but included 12 stamped amphora handles (Rose, pers. comm.). The contrast with Memphis requires an explanation. This is probably that the workmen's village, with its regular plan, housed a workforce engaged upon official projects and in part supplied from the royal (or temple) storerooms, though whether or not the amphorae arrived in the village with their original contents remains unknown. The complexes of mud brick houses at Kom Rabi'a, on the other hand, show intricate building sequences with every appearance of having evolved according to need, and the inhabitants of such a district are unlikely to have had access to regular supplies from any official source, (though for a different view see Giddy 1999, 3).

So amphorae of a characteristically Egyptian form seem to have appeared around the beginning of the 18th Dynasty. To try to discover why this occurred, we must consider their Middle Kingdom and Second Intermediate Period predecessors. Firstly, can we say what type of vessel was used at that time for the transportation of liquids such as wine? Secondly, if (as is generally agreed) the earliest Egyptian amphorae were inspired by imported Canaanite jars of the Middle Bronze period, which were common in the eastern Delta and the Memphite region (Bourriau 1990; Arnold, Arnold and Allen 1995; McGovern 2000), why did inspiration not strike earlier than it did?

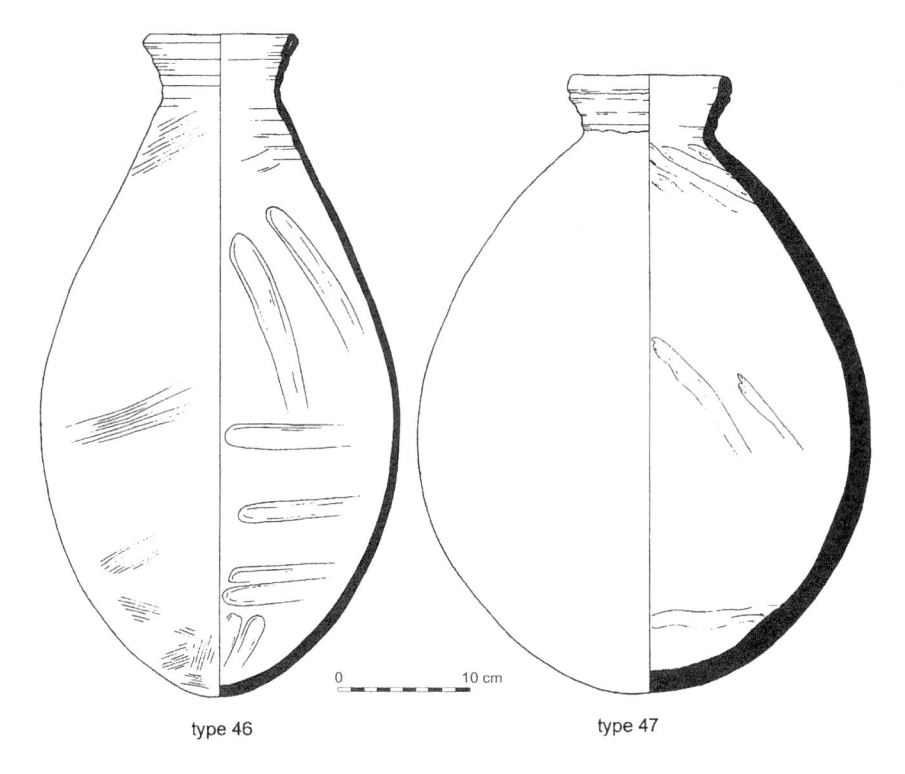

Figure 5.5 Corrugated necked jars of Marl C from Tell el-Dab'a (Bader 2002, Fig. 7, types 46–7).

The commonest fabric of Egyptian transport and storage jars at this time was Marl C (Bader 2001), the only marl fabric utilised for this purpose in Lower Egypt. The relatively few vessels of other marl fabrics found in Lower Egypt derive from elsewhere, mainly the Theban region (Arnold 1981; Nordström and Bourriau 1993, 179–80). There is one jar type (Figure 5.5) (Bader 2001,129–146; Bader 2002, 41–3) which, on account of its capacity and wide distribution outside the Memphite region (in the Delta, Middle and Upper Egypt and Nubia), as well as the occurrence of the hieratic inscription "sweet wine of Syria" on one example, we may suppose to have been used for the transport of wine, though perhaps also of other commodities. These jars were enormously heavy because they were made by hand coiling, and, despite assiduous finger smoothing, their walls are very thick. They had no handles, so the corrugated neck was not decorative but functional, intended to receive a carrying rope or the drawstring of a bag. Traces of fibre have been found around the neck of one example (Susan Allen pers.comm.). To carry a full jar would have required two men with a thick, heavy pole, though a donkey could have managed two, one in each pannier. They could have been transported by boat, but the full-bottomed shape of the Marl C jar, unlike that of its Canaanite counterpart, does not make for efficient stacking.

We can now turn our attention to the commodity we assume to have been transported, namely wine. There is no doubt that the principal vine-growing areas of Egypt were then, as in the New Kingdom, the Delta and the Western Oases, but there is no evidence that it was on any very large scale (Meeks 1993; James 1996; Murray 2000). On his stela Kamose, the King of Thebes (c. 1550 B.C.), describes a raid on the Eastern Delta; before the citadel at Avaris (Tell el Dab'a) he boasts: " I will drink the wine of his (the Hyksos King Apophis') vineyard, which the Asiatics whom I captured, press for me." This tells us that wine had high status (Kamose would hardly boast of drinking the King's beer!), that it was a speciality of the foreigners of Avaris, and that the technology of wine-making was not familiar at Thebes at that time. Athough there is supporting physical evidence from Avaris (Tell el Dab'a) of the existence of vines, there is also an extraordinary number of Canaanite Jars from the site, and this is how, I judge, most of the wine consumed there was obtained. In his recent study (McGovern 2000) McGovern has confirmed by residue analysis that the commodity in these imported jars was wine and by Neutron Activation analysis of the clays suggested that they originated in the city-states of Southern Palestine.

Whatever the provenances of the imported wine in the Middle Kingdom and Second Intermediate Period, at the beginning of the 18th Dynasty the scenario totally changed. Kamose's successor, Ahmose, finally drove out the Hyksos King and firmly established himself at Avaris, building a great palace. Egypt was unified and the fertile Delta and its products could be sent south without hindrance while the Hyksos' trade in wine was interrrupted. Recent studies of Canaanite Jars from Egypt in the New Kingdom (Bourriau, Smith and Serpico 2001; Serpico *et al.* 2003; Smith *et al.*, this volume) have shown that from the beginning of the 18th Dynasty onwards the commodities carried in the jars had changed. The products imported to Memphis and Amarna in the New Kingdom were oils and resin rather than wine and the provenance of most of the jars, as indicated by the fabrics, was the coast of northern Israel, Lebanon and Syria; none could be shown to come from the area suggested by McGovern for the imports of the preceding Middle Bronze period (McGovern 2000).

I suggest that one of the measures which Ahmose and his successsors undertook was to increase local production of wine to replace the imported varieties. This entailed extending the existing Delta vineyards and placing the whole organisation under royal bureaucratic control. That first royal stamp of Tuthmosis I can be dated approximately 25 years after the conquest of Avaris by Ahmose. In this context, it is interesting to note the discovery at Tell Dab'a of a vinyard and wine press, dated by Bietak to the early Eighteenth Dynasty (Bietak 1985).

Such royal encouragement of viticulture in the Delta at the beginning of the 18th Dynasty suddenly required transport amphorae and this brings us back to the subject of clay since the only clay available in the Delta was alluvial silt, unsuitable, because of its porosity, for the long-term storage of liquids. An interesting experiment (Hope 1989,109) tested the porosity of a Marl D amphora, this being the fabric of the early amphorae from Memphis, and found it to be quite efficient for the purpose. Hope's example was slipped, unlike the one from Level V at Memphis, and this would have reduced its permeability. The vessel held 17.3 litres when full and weighed 21 kilos. The volume of water lost after 11 days was 0.675 and after 21 days 1.45 litres. Most of the loss occurred where the slip had worn away.

Seen macroscopically (Nordström and Bourriau 1993, Pl.VII, a–c, e–f) Marl D is dense and hard and usually shows narrow red outer zones and a grey or grey-brown core on the fresh sherd break. Inclusions are fine quartz (common); fine to medium limestone (common); occasional shell and foramanifera; occasional fine mica; and occasional soft red-brown fine to medium particles of ochre (Bourriau and Nicholson 1992, H1). The fabric is not homogeneous: it is a fabric group rather than a single fabric. At Memphis two variants were identified (H1 and H14) and at Amarna four (III.2; III.3; III.5; III.6). Petrological examination by Paul Nicholson suggests that we are dealing with marl clays which are similar but originating from different limestone sources. He notes in the Memphite samples a distinction between a group with high quartz and fossils and a coarser group with reduced quartz and no fossils. The Amarna fabrics in the group each show lower standard variations than the Memphis fabrics but this may be explained in two ways: firstly, the samples have come from contexts with a much shorter time range, about 25 years as opposed to about 300 years at Memphis, and secondly, the ceramicist, Pamela Rose, has made more subdivisions within the fabric group than has been done at Memphis.

Much work has recently been done on Egyptian limestone quarries and 88 have been mapped between Esna and Giza (Aston, Harrell and Shaw 2000 and bibliography there cited). Of course, we cannot assume that such quarries were also the sources of potters' clay but the evidence does allow us to characterise the limestone from different parts of Egypt. The closest quarry to the Delta is at Mallahet Mariut, just west of Alexandria – a region certainly exploited for industrial production of amphorae in the Ptolemaic and Roman period, though there is at present no evidence that it was exploited any earlier.

It may be possible eventually to combine the evidence from ceramic thin sections with that from a reference collection of thin sections from the quarries and suggest sources for Marl D, at least on a regional basis. There are formidable difficulties to overcome, firstly in identifying, petrographically, the changes in the raw material wrought by the potter's processing and firing, and secondly, in the recognition in the ceramic thin section of fossils or mineral constituents distinctive of a particular limestone.

For the present, the only published analytical study focusing on the provenance of Marl D is that of McGovern (1997). On the evidence of Neutron Activation Analysis of samples from Malqata, Thebes, followed by statistical comparison with a large databank of Egyptian clay and fabric samples and a control group of samples supposedly of Theban origin, he postulated a Theban source for Marl D. All the archaeological evidence had suggested an origin in the Memphite region and the petrological study had suggested more than one source for the fabric group. McGovern's suggested provenance is open to criticism in respect of his control group. The control group of seven sample sherds chosen to represent clays of Theban origin were all *found* in Theban excavations, but two were of Marl C, for which McGovern has himself suggested a source close to the Faiyum (McGovern 1997, 95); three came from Marl D amphora (so of the same fabric and date as the samples whose provenance was in question but offering no additional evidence that they were manufactured in Thebes) and two others were of fabrics, Marl A2 and Marl B, thought to be of Theban origin on archaeological grounds but which other NAA data (see below) had clearly separated from a group of Marl D samples.

This second NAA study, soon to be published (Bourriau *et al.* in press), set out to test the fabric groupings of the Vienna System as a whole. Thirty seven of the 351 samples analysed were visually identified as belonging to the Marl D fabric group before the analysis. Fifteen of these grouped by NAA with the Nile clay samples and a subsequent study (Bourriau, Smith and Nicholson, 2000) confirmed that they originated from Nile clays or a mixture of Nile and Marl clays. Of the remaining twenty two samples, six were outliers within both the Nile and Marl groups, and one, a very uncertain Marl D, was later re-identified as a Canaanite amphora fabric. However, fifteen Marl D samples fell into a tight

	MD MEAN	SD
Na %	.505	.244
Al %	6.29	.698
Ca %	14.80	2.29
Sc ppm	15.5	.610
Ti %	.519	.0490
V ppm	124	9.03
Cr pm	117	8.31
Mn ppm	878	148
Fe %	4.75	.195
Co ppm	20.0	.976
Rb ppm	39.2	3.93
Cs ppm	1.85	.286
La ppm	32.7	2.32
Ce ppm	67.1	3.71
Sm ppm	6.11	.466
Eu ppm	1.35	.0742
Dy ppm	3.86	.303
Lu ppm	.391	.0782
Hf ppm	5.93	.416
Ta ppm	1.45	.147
Th ppm	7.23	.442
U ppm	2.3	.395

Figure 5.6 Means and standard deviations of element concentrations for 15 samples of Marl D, analysed by Neutron Activation Analysis (Bourriau et al. forthcoming).

88   *Janine Bourriau*

chemical group which contained only two other samples (one was identified visually as Marl A2 and the other as Marl A3). The table in Figure 5.6 shows the composition of this group, giving the average ('mean') value of each element and the standard deviation between the samples. The sherds were found at Thebes (10), Qantir (near Tell el Dab'a) (3), Amarna (2), Memphis (1) and Armant (in Upper Egypt) (1). This result does not confirm either a Theban or a Memphite origin, but does strongly suggest a single source for the clay, and one in use over a long period of time, since the samples ranged in date from c. 1350 to c. 1225 B.C. It also confirms that Marl D has a distinct chemical fingerprint (confirmed by McGovern's data also) which differentiates it from other Marl clays of the New Kingdom, such as Marl A4, and Marl B. It should be noted that this group of Marl D samples included two from Amarna but the NAA data did not, unlike the petrological data, distinguish them.

These considerations, though they do not disprove a Theban origin for Marl D, and thus of the wine amphorae made of this clay, do cast doubt upon it and thus upon McGovern's theory of a trade in empty amphorae made in Thebes sent to the Delta to be filled, labelled and returned to Thebes for the many royal festivals which took place there. A source for Marl D in the Memphite region close to the Delta and Faiyum vineyards, still seems the most likely and is supported by the early date of the amphora from Kom Rabi'a, Memphis.

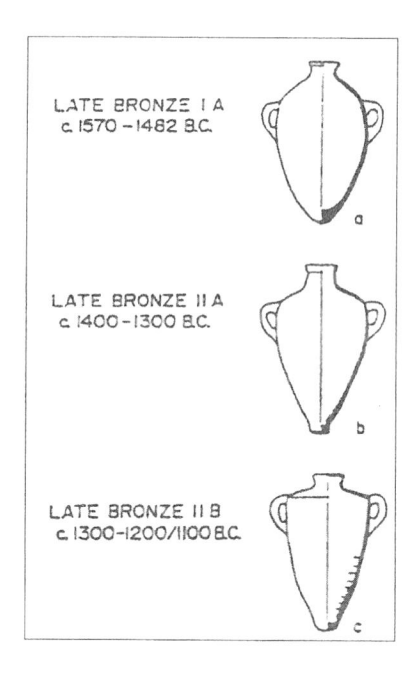

Figure 5.7 The development of the Canaanite Jar in the Late Bronze Age (Leonard 1996, Figs. 15.2, 15.3).

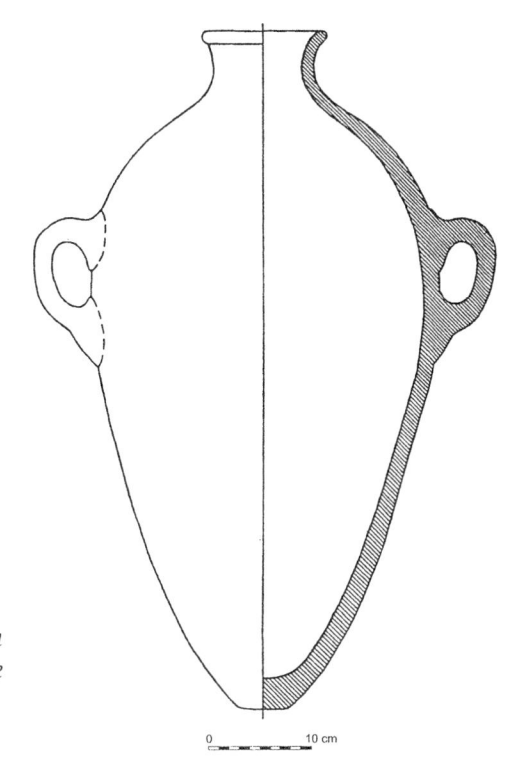

Figure 5.8 Amphora of Queen Meryet-Amun, probably Canaanite (Winlock 1932, Fig. 17).

In the light of this new information, it is necessary to look again at some earlier discussions of Egyptian amphorae, especially Wood (1987) and Leonard (1996). The close similarity of shape of the Late Bronze Age I transport amphora and the Egyptian amphora is undeniable (Figure 5.7) but the term 'Canaanite Jar' or 'Egyptian Canaanite Jar' should not be perpetuated, since it obscures the boundary between two distinct classes: one is a true Canaanite Jar, made in a variety of clays and used to import from various regions of the Levant into Egypt a whole range of commodities – oils, resin and honey, as well as wine – while the other is an Egyptian product with a different base-forming technology, made almost exclusively to carry wine (but note Murray's caveat, 2000, 580) within Egypt and Nubia. It is clearly of the utmost importance not to confuse the two. Moreover, Leonard's supposition that Canaanite Jars always carried wine to Egypt and the Aegean, and can serve in the New Kingdom as an indicator of a Levant-wide wine trade, has to be abandoned.

Where the fabric cannot be ascertained, despite the similarity of shape, it may sometimes be possible to determine on shape properties alone whether or not a vessel is an import. Two examples from the early 18th Dynasty illustrate this. The first is the amphora from the burial of Queen Meryet-Amun. This should be a

Canaanite Jar, probably of our Group 1 (see Smith, this volume) judging by the fabric description (Winlock 1932, 31–2, n.9, 74) as "red pottery with a drab slip of *kulleh* clay" (although such a description would also suit Marl D with a slip). It held a beer residue, so was clearly re-used. The position of the handles, thickness of vessel wall and (apparently) mould-made (or turned, Rye 1981,134–7) base, as opposed to the coiled base of the Memphis amphora, all indicate a Canaanite, rather than an Egyptian amphora.

The second example is from an intact surface burial excavated by Emery at Saqqara, closely dateable by its ceramics to the reign of Amenophis I (1525–1504 B.C.) (Bourriau 1991a, Fig.6). The whole group is in the British Museum, except for the amphora, the whereabouts of which are unknown (Fig.9, 14). The excavator recorded the contents as being milk (though without giving the evidence), so we can only assume it to be another example of re-use. The maximum diameter lies close to the vessel's mid-point, the vessel wall is very thick and the handles lie lower (like those of Meryet-Amun's amphora) on the profile than those on the Memphis amphora. I judge this to have been another Canaanite amphora. The elongated lower body of the Egyptian amphora shape, visible in the amphora of Nakht Min (Figure 5.3), became even more marked in the later 18th Dynasty examples. Egyptian amphorae also retained a rounded shoulder, whereas by the late 18th Dynasty the Canaanite Jar had evolved a carinated shoulder (Figure 5.7) which was to remain a characteristic feature for hundreds of years.

New Kingdom amphorae represent an important advance compared with the transport vessels of the Middle Kingdom. Because they were wheel-thrown, and made in sections, the vessel wall could be as thin as 5 mm. As a result, each filled amphora was much lighter and (unlike a Marl C jar) could be lifted and carried on the shoulders of one man.

The reunification of Egypt under a Theban family at the beginning of the 18th Dynasty, took place after a long period during which the Nile below Hermopolis had probably been closed to Upper Egyptians (Bourriau 2000). The Delta and its products were newly accessible, and this included the royal vineyards, and experienced Asiatic vintners. Even later in the New Kingdom, vintners often bore non-Egyptian Semitic names: two of the fifteen named on the jars from the tomb of Tutankhamun had Syrian names (Lesko 1996). In fact, wine-making in the Delta may begin much earlier: the earliest evidence of grapes comes from Buto in the central and Tell Ibrahim Awad in the eastern Delta (Murray 2000 and bibliography cited there), though this is not quite evidence of viticulture.

Wine production on a large scale presupposes the production of jars on a comparable scale. The standardised shape, fabric and method of manufacture of these jars (which become even clearer later in the Dynasty) suggest specialised workshops (Bourriau, Smith and Nicholson 2000). Within 25 years of the defeat of the Hyksos, royal control is evident – royal control which is, in an Egyptian context, essential. In the New Kingdom, the King's offering of wine to the gods became a significant part of royal ritual (Mu-Chou Poo 1995). Wine became an important indicator of status and a measure of royal patronage (Tallet 1998). An inscription of Ramesses III records the offering of 59,588 jars of wine to Amun,

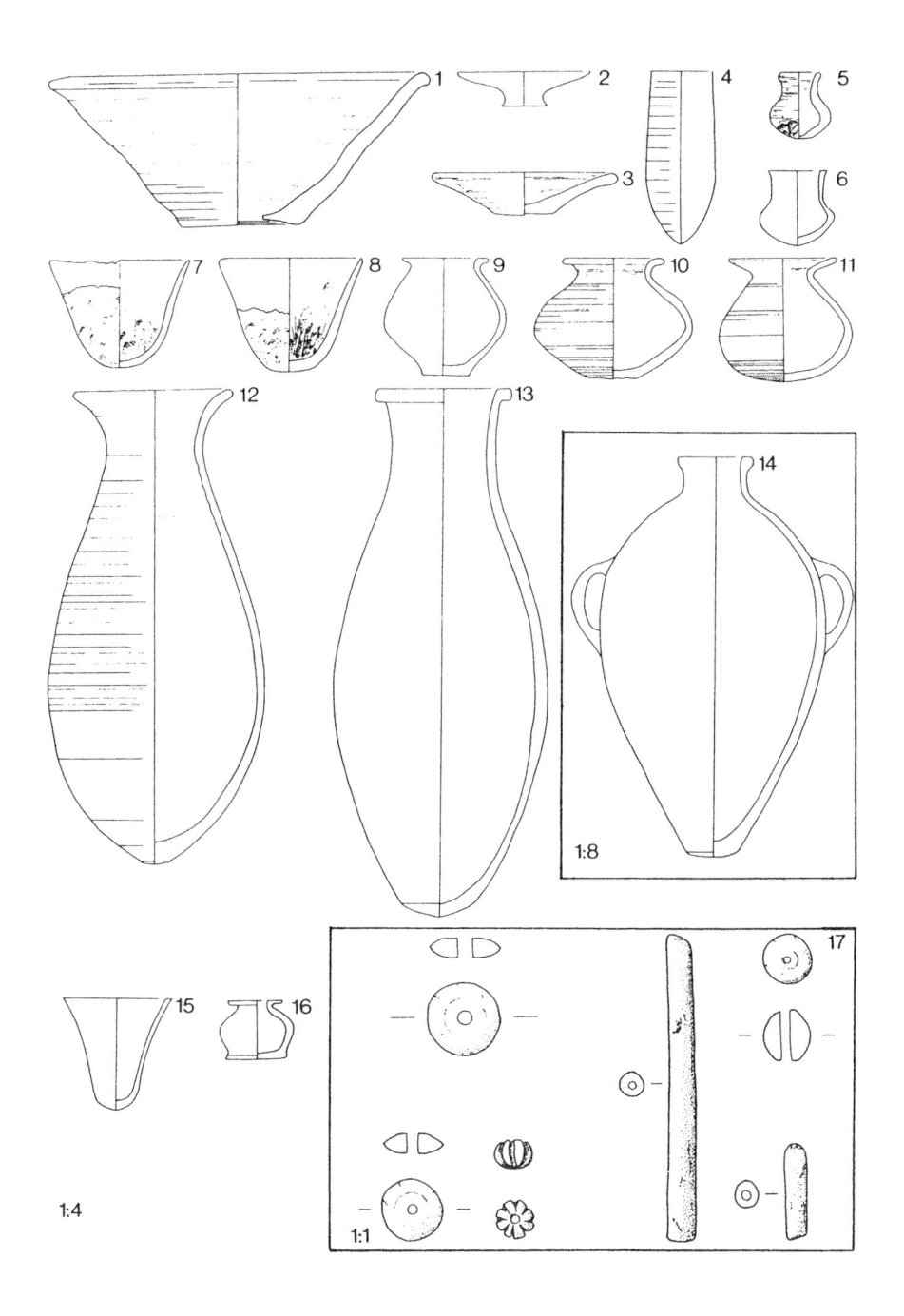

Figure 5.9 Amphora (No. 14), pottery, stone vessels and beads from an intact burial at Saqqara (Bourriau 1991a, Fig. 6).

and the Harris Papyrus the same King's gift of 433 vineyards and groves to the Theban temples alone.

There were also private vineyards in the Delta, at least as early as the reign of Tuthmosis III (1479–1425 B.C.) and probably much earlier. The Theban tomb of the Royal Herald Intef (Lesko 1996, Figs.14.2, 14.3) has scenes of grape pressing, and of the storage of sealed wine amphorae recognisably of the type of the contemporary amphora of Nakht Min. Such vineyards would have increased even further the demand for amphorae.

We should now return to the evidence from New Kingdom Level IV at Memphis. It is striking how little Marl D is to be found in this level, which is thought at present to cover the 18th Dynasty until the end of the reign of Tuthmosis III. Marl D is heavily outweighed by Marl A2 and Marl B, the fabrics of storage jars imported from Thebes, where they have a long previous history, going back to the Second Intermediate Period. We may speculate that Marl D amphorae, as an innovation, were working their way gradually down through society, starting at the top. The few amphorae which found their way to Kom Rabi'a were probably being re-used, by analogy with the use of Egyptian and imported amphorae to bring well-water to the workmen's village at Amarna (Renfrew 1987).

Similarly, even rich élite burials do not contain Marl D amphorae prior to the reign of Hatshepsut (1473–1458 B.C.), which appears to indicate that the distribution of wine from the royal storerooms to the tombs of officials, such as Nakht Min, was being carried out in accordance with a strict system of rotation. The system of exchange, gift and purchase which determined the circulation of the wine amphorae has recently been unravelled from the evidence of the labels (Tallet 1998).

In conclusion, I should like to suggest that although the association of Egyptian wine and Egyptian amphorae continues for a very long time, it eventually languishes. The suggestion which follows remains to be tested against the post-New Kingdom written sources so usefully summarised by Dimitri Meeks (1993), but in the late Third Intermediate Period at Buto I have identified at least seven Canaanite transport amphora fabrics in addition to that of the ubiquitous Phoenician wine amphora; and only one Egyptian amphora class. The imports overwhelmingly outweigh the local amphorae. If this reflects, as it seems, the proportion of imported to local wine consumed, it recalls the often-quoted remark of Herodotus, writing in the slightly later Persian period, that "(the Egyptians) drink wine made from barley as they have no vines in the country" (Herodotus, 131). Only much later, in the 3rd century B.C., did the deliberate encouragement of wine production in the Delta and the Faiyum by the Ptolemies rejuvenate the Egyptian amphora industry, and now the prototype was the Greek transport amphora, not the Canaanite Jar.

REFERENCES

Arnold, D., Arnold, F. and Allen, S., 1995, Canaanite Imports at Lisht, the Middle Kingdom Capital of Egypt, *Ägypten und Levante* V, 13–32.

Arnold, D., 1981, Ägyptische Mergeltone ('Wüstentone') und die Herkunft einer Mergeltonware des Mittleren Reiches aus der Gegend von Memphis, in D. A r n o l d (ed.), *Studien zur altägyptischen Keramik*, 167–91. Von Zabern, Mainz.

Aston, D., Harrell, J. and Shaw, I., 2000, Stone, in P. T. Nicholson and I. Shaw (eds.), *Ancient Egyptian Materials and Technology*, 5–77. Cambridge University Press, Cambridge.

Bader, B., 2001, *Tell el-Dab'a XIII. Typologie und Chronologie der Mergel C-Ton Keramik. Materialien zum Binnenhandel des Mittleren Reiches und der Zweiten Zwischenzeit.* Österreichische Akademie der Wissenschaften, Vienna.

Bader, B., 2002, A Concise Guide to Marl C Pottery, *Ägypten und Levante*, XII, 29–54.

Bavay, L., Marchand, S. and Tallet, P., 2000, Les jarres inscrites du Nouvel Empire provenant de Deir el-Médineh, *Cahiers de la Céramique Égyptienne*, 6, 77–89.

Bietak, M., 1985, Ein altägyptische Weingarten in einem Tempelbezirk, *Anzeiger der Phil.-hist. Klasse der Psterreichischen Akademie der Wissenschaften* 122, 267–278.

Bourriau, J. D., 1990, Canaanite Jars from New Kingdom Deposits at Memphis, Kom Rabi a, *Eretz-Israel*, 21, 18–26.

Bourriau, J. D., 1991a, Relations between Egypt and Kerma during the Middle and New Kingdoms, in W.V.Davies (ed.), *Egypt and Africa. Nubia from Prehistory to Islam*, 129–44. British Museum Press, London.

Bourriau, J. D., 1991b. The Memphis Pottery Project, *The Cambridge Archaeological Journal*, 2, 263–268.

Bourriau, J. D., 1997, Beyond Avaris: The Second Intermediate Period in Egypt outside the Eastern Delta, in E. D. Oren (ed.), *The Hyksos: New Historical and Archaeological Perspectives*, 159–182. University of Pennsylvania Museum of Archaeology and Anthropology, Philadelphia.

Bourriau, J. D., 2000, The Second Intermediate Period, in I. Shaw (ed.), *The Oxford History of Ancient Egypt*, 185–217, 461–2. Oxford University Press, Oxford.

Bourriau, J. D. and Nicholson, P. T., 1992, Marl Clay Pottery Fabrics of the New Kingdom from Memphis, Saqqara and Amarna, *Journal of Egyptian Archaeology*, 78, 29–91.

Bourriau, J. D. and Eriksson, K. O., 1997, A Late Minoan Sherd from an Early 18th Dynasty Context at Kom Rabi a, Memphis, in J. Phillips (ed.), *Ancient Egypt, the Aegean, and the Near East: Studies in Honor of Martha Rhoads Bell*, 95–120. Van Siclen Books, St. Antonio, Texas.

Bourriau, J. D., Smith, L. M. V. and Nicholson, P. T., 2000, *New Kingdom Pottery Fabrics: Nile clay and mixed Nile/Marl clay fabrics from Memphis and Amarna.* Egypt Exploration Society, London.

Bourriau, J. D., Smith, L. M. V. and Serpico, M., 2001, The Provenance of Canaanite Amphorae found at Memphis and Amarna in the New Kingdom, in A. Shortland (ed.), *The Social Context of Technological Change: Egypt and the Near East, 1650–1150 B. C.*, 113–146. Oxbow, Oxford.

Bourriau, J. D., Bellido, A., Bryan, N. and Robinson, V.†, (forthcoming), Egyptian Pottery Fabrics: A Comparison between Chemical Groups using NAA and the "Vienna System".

Guksch, H., 1995, *Die Gräber des Nacht-Min und des Men-cheper-Ra-seneb. Theben Nr. 87 und 79.* Von Zabern, Mainz.

Herodotus, 1954, *The Histories*, trans. A. de Sélincourt. Penguin Classics, London.

Hope, C. A., 1978, *Malkata. Jar Sealings and Amphorae*. Aris and Phillips, Warminster.

Hope, C. A., 1989, *Pottery of the Egyptian New Kingdom. Three Studies.* Victoria College Press, Victoria, Australia.

Hope, C.A., 2002, Oases amphorae of the New Kingdom in R.Friedman (ed.), *Egypt and Nubia. Gifts of the Desert*, 95–131. British Museum Press, London.

Hope, C. A., Blauer, H. M. and Riederer, J., 1981, Recent Analyses of 18th Dynasty Pottery, in D. Arnold (ed.), *Studien zur altägyptischen Keramik*,139–166. Von Zabern, Mainz.

Hughes, G. R., 1963, Serra East: The University of Chicago Excavations, 1961–2. A Preliminary Report on the First Season's Work, in *Kush* XI, 121–130.

Giddy, L., 1999, *Kom Rabi a: the New Kingdom and Post-New Kingdom Objects. Survey of Memphis II.* Egypt Exploration Society. London.

James, T. G. H., 1996, The Earliest History of Wine and Its Importance in Ancient Egypt, in P. E. McGovern, S. J. Fleming and S. H. Katz (eds.), *The Origins and Ancient History of Wine*, 197–213. University of Pennsylvania Museum of Archaeology and Anthropology, Philadelphia.

Koenig, I.,1979, *Catalogue des étiquettes de jarres hiératiques de Deir el-Médineh I.* Institut Français d'Archéologie Orientale, Cairo.

Leonard, A., 1996, "Canaanite Jars" and the Late Bronze Age Aegeo-Levantine Wine Trade, in P. E. McGovern, S. J. Fleming and S. H. Katz (eds.), *The Origins and Ancient History of Wine*, 233–254. University of Pennsylvania Museum of Archaeology and Anthropology, Philadelphia.

Lesko, L., 1977, *King Tut's Wine Cellar*. B. C. Scribe Publications, Berkeley.

Lesko, L., 1996, Egyptian Wine Production during the New Kingdom, in P. E. McGovern, S. J. Fleming and S. H. Katz (eds.), *The Origins and Ancient History of Wine*, 215–230. University of Pennsylvania Museum of Archaeology and Anthropology, Philadelphia.

McGovern, P. E., 1997, Wine of Egypt's Golden Age, *Journal of Egyptian Archaeology* 83, 69–108.

McGovern, P. E., 2000, *The Foreign Relations of the "Hyksos". A Neutron Activation Study of Middle Bronze Age Pottery from the Eastern Mediterranean.* BAR International Series 888. Archaeopress, Oxford.

McGovern, S. J. Fleming and S. H. Katz (eds.), *The Origins and Ancient History of Wine.* University of Pennsylvania Museum of Archaeology and Anthropology, Philadelphia.

Meeks, D., 1993, Oléiculture et Viticulture dans l'Égypte Pharaonique, in M-C.Amouretti and J-P. Brun (eds.), La production du vin et de l'huile en Méditerranée, 3–38. *Bulletin de Correspondance Hellénique. Supplément XXVI. École Française d'Athènes*, Athens.

Mu-Chou Poo, 1995, *Wine and Wine Offering in the Religion of Ancient Egypt.* Kegan Paul International, London.

Murray, M. A., 2000, Viticulture and Wine Production, in P. T. Nicholson and I. Shaw (eds.), *Ancient Egyptian Materials and Technology,* 577–599.Cambridge University Press, Cambridge.

Nicholson, P. T. and Rose, P. R., 1985, Pottery fabrics and Ware groups at el-Amarna, in B. J. Kemp, *Amarna Reports II*, 133–174. Egypt Exploration Society, London.

Nordström, H-Å. and Bourriau, J. D., 1993, Ceramic Technology: Clays and Fabrics, in D. Arnold and J. D. Bourriau (eds.), *An Introduction to Ancient Egyptian Pottery.* Von Zabern, Mainz.

Renfrew, C., 1987, Survey of Site X2, in B. J. Kemp, *Amarna Reports IV*, 87–102. Egypt Exploration Society, London.

Rose, P.R., 1984, Pottery distribution analysis, in B. J. Kemp, *Amarna Reports I*, 133–153. Egypt Exploration Society, London.

Rye, O. S., 1981. *Pottery Technology. Principles and Reconstruction*. Taraxacum, Washington.

Serpico, M., Bourriau, J., Smith, L., Goren, Y., Stern B. and Heron, C. 2003. Commodities and Containers; A Project to Study Canaanite Amphorae Imported into Egypt during the New Kingdom, in M. Bietak (ed.), *The Synchronisation of Civilisations in the Eastern Mediterranean in the Second Millennium B.C. II*, 365–375. Österreichische Akademie der Wissenschaften, Vienna.

Tallet, P., 1998, Quelques Aspects de l'Économie du Vin en Égypte Ancienne au Nouvel Empire, in N. Grimal and B. Menu (eds.) *Le commerce en Égypte ancienne*. Institut Français d'Archéologie Orientale, Cairo.

Winlock, H. E., 1932, *The Tomb of Queen Meryet-Amun at Thebes*. Metropolitan Museum of Art, New York.

Wood, B.G., 1987, Egyptian Amphorae of the New Kingdom and Ramesside Periods, *Biblical Archaeologist* 50, 75–83.

Chapter 6

Natural Product Technology in New Kingdom Egypt

Margaret Serpico

Abstract

Although ancient inorganic materials such as glass and metal have been widely studied, a range of organic natural products was also important in daily life. However, in contrast to the research into the production of inorganic remains, attempts to provide a general comparative overview of the technologies used to collect and manufacture natural products have been largely absent. Within the context of presenting a synthesized study, changes in the sources and technology during the Late Bronze Age can be examined. Recent examples of scientific analyses can also be integrated. To place this subject into a broader context, the titles and social status held by people working in this industry are briefly considered, and also the impact that such a high demand for these products might have had on political events.

INTRODUCTION

Within the collective group of organic materials known in the ancient world is a subset which has been termed 'natural products.' These substances (Figure 6.1), such as oils, fats, wax, honey, gum, resin, amber and bitumen, can be defined as organic materials that, unlike bone, wood, or plant remains for example, lack a recognizable macroscopic, cellular structure. They can only be securely identified through analysis of their characteristic chemical constituents, often termed biomarkers, most frequently by the scientific technique of gas chromatography/mass spectrometry (GC/MS) (Evans and Heron 1993, 446). These products were often used for a range of purposes spanning both secular and religious contexts and including ritual, medicinal and utilitarian functions (Figure 6.2). Generally excluded as another organic subset are the vast array of possible food products including beer and wine, although some natural products could have been used for this purpose.

Common Natural Products

Oil	Fat	Wax	Honey
Resin	Fossil Resin	Gum	Bitumen

Figure 6.1 List of natural products available to the ancient Egyptians.

Selection of Possible Uses
(individually and in mixtures)

Adhesives	Foods	Cosmetic Preparations
Illuminants	Emollients	Censing
Preservatives	Insect repellent	Medicaments
Sealants	Cleansers	Ritual Preparations
Lubricants	Fumigants	Magical Preparations

Figure 6.2 List of some of the possible uses of natural products in ancient Egypt.

Clearly, given its scientific basis, the term 'natural product' is an artificial one. While the term is undoubtedly a useful description of these substances, it is also in itself confusing. Raw materials such as resin, gum, amber and bitumen are classed as natural products, while other substances usually included in this group require preparation and treatment to produce the final product. Moreover, there is little evidence that the Egyptians generally viewed these substances collectively, although administrative records of the New Kingdom sometimes list a single entry for a number of pottery jars which together held products such as oil, resin and honey (*e.g.*, Sethe 1907, 688 [8]; Erichsen 1933, 37, (32b, 3; 33b, 8); Grandet 1994, 267, 269). Similarly, many of these ancient products, either individually or in mixtures, are preserved today as deposits in pottery, glass and stone jars. Although it is tempting to look at jar residues and make a visual identification of the likely contents, such as a resin or an oil, these assessments are especially risky. They can enter into the literature and assume an authority that is, in fact, unsupported. Moreover, the contents are seldom appreciated as the result of a range of industrial processes. To some extent, this is understandable given the lack of clear terminology or knowledge of the stages of production, such as exists for many of the inorganic industries, for example, metallurgy.

As 'natural products' are an artificial group, it is difficult to unify the procedures used, but because these commodities were often mixed, usually stored in stone, glass or pottery jars, and are most successfully distinguished by the same chemical techniques, there is some merit in attempting to consolidate the manufacturing stages. Such a systematic approach can highlight the production similarities and differences between the natural products, and also facilitate comparisons to inorganic products. In addition, it can enable a better understanding of the comparative social and administrative aspects of production by relating the different stages of manufacture to certain occupations and/or levels of social status.

Figures 6.3 and 6.4 are basic diagrams which attempt to consolidate the possible sequences for processing natural products. From these diagrams, it is evident that even when jar contents can be successfully identified through scientific

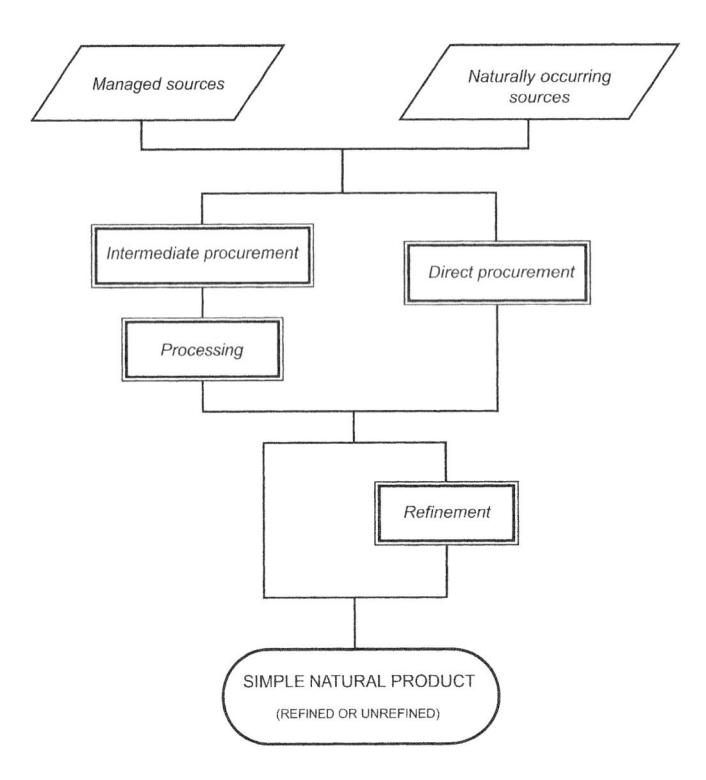

Figure 6.3 Possible stages in the production of simple natural products.

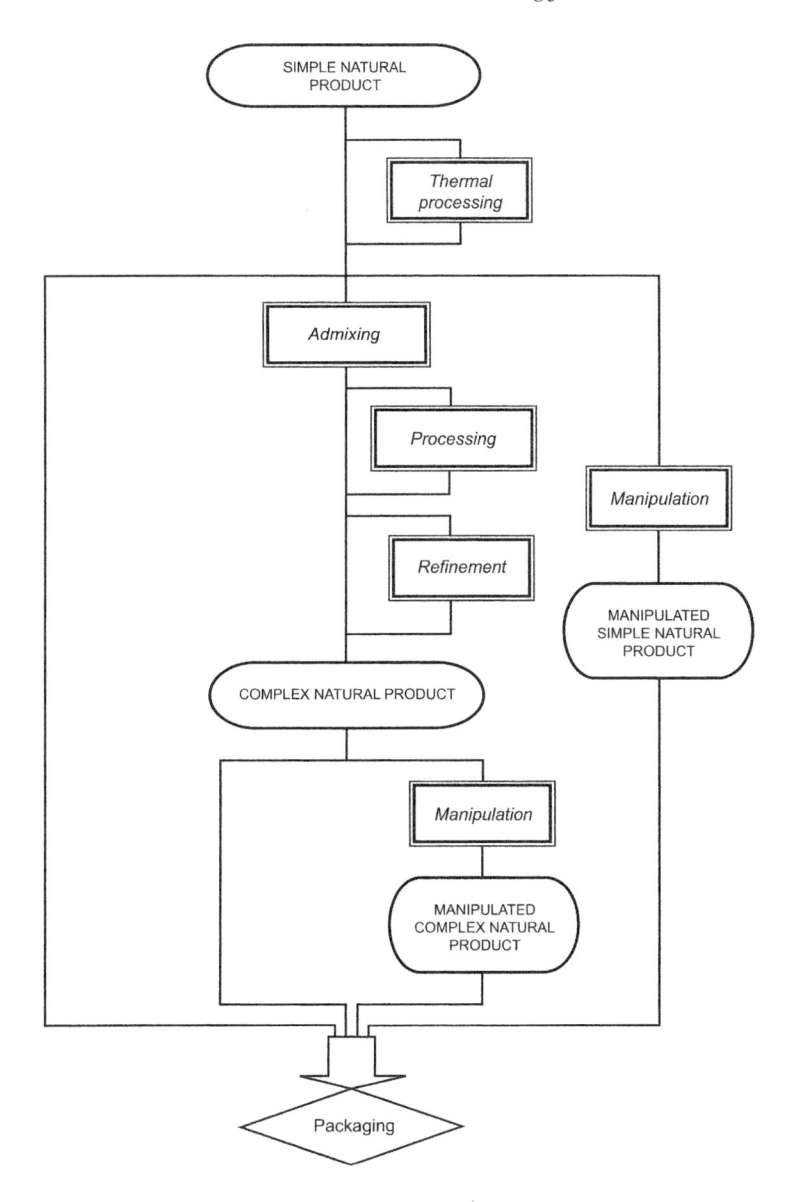

Figure 6.4 Possible stages in the production of complex natural products.

analysis, the final product could be the result of a number of different technological pathways. Other concerns include the possibility that the contents of jars represent reuse or that other components, now undetectable, were once present in the contents.

SIMPLE NATURAL PRODUCTS

Sources

The first distinction that can be made is between natural products and natural product sources. Sources include oil plants, animals, beehives, gummy and resinous plants and amber and bituminous outcrops. These can be divided into those which occur spontaneously or are found wild and those which result from some type of human management. Examples of the former are wild gummy and resinous plants and discreet deposits of amber and bitumen. Although they can be tapped, gums and resins will produce exudates whenever the source plant is injured through damage or insect attack and, like bitumen and amber, this can occur without human intervention. These sources can be collected as raw, simple natural products and then stored for use without any further processing.

Other sources are more difficult to classify as they can cross over both categories. Oil plants can be cultivated or, in some cases, collected wild; animal fats can be obtained from wild animals or domesticated sources; honey and beeswax can be extracted from wild hives or from a managed apiary. The distinction between naturally occurring and domesticated sources is important as the latter represent a high level of organization, time investment and commitment to the production. Many more stages are involved (planting and watering of crops, rearing of animals, building of beehives, for example) to arrive at the point of procuring the raw source material. Moreover, some limited forms of management may also have been undertaken on naturally occurring sources, such as controlling density of plants and removal of unwanted specimens, or the pruning of trees.

For the period under consideration, it is important to recognize that some natural product sources did not occur in Egypt and that other new sources were introduced. General overviews of the range of natural product sources are available (Lucas 1962, 1–7, 80–97, 303–337; Murray 2000, 614, Table 24.1; Serpico and White 2000b, 390–429; Serpico 2000, 430–474; Newman and Serpico 2000, 475–494), so the focus here is on changes in the availability of those sources during the New Kingdom.

Notably, Egypt has very limited supplies of resins, and the extent to which internal sources were exploited is unclear. As a result, it is not yet possible to determine any definite changes in the range of botanical sources for resins in the New Kingdom. The extensive use of *Pistacia* sp. resin during this time has been established through GC/MS analysis of resin samples, and it is clear that this resin was known as *sntr* during the New Kingdom and widely used as incense (Serpico and White 2000a; Serpico and White 2001; Stern *et al.* 2003). However, the use of incense and occurrences of the word *sntr* date back to the early dynastic times and there has not so far been sufficient analytical research on resin samples of earlier date to put this into the necessary chronological context. It is possible that the word designated other resin sources depending on the time period or location where the resin was obtained.

Gum-resins such as frankincense and myrrh, would both have had to have been imported. It is doubtful that the Egyptians ever successfully introduced or cultivated imported resin or gum-resin plants, but one possible attempt is shown in the reliefs from the temple of Hatshepsut, where trees of `ntyw, believed to be frankincense or myrrh, were brought back to Egypt and planted at her temple at Deir el Bahari (Naville 1898, Pls. LXIX–LXXVI). However, it appears that the endeavor failed and no further attempts were made (Dixon 1969; Hepper 1969).

Still less is known about changes of sources of gum or amber during the New Kingdom. Locally available gum sources include *Acacia* and some species of *Astragalus*, although sources from other locations, particularly Sudan, may also have been available (Newman and Serpico 2000, 475–80). With regard to amber, Baltic specimens are well known, but varieties can also be found in a number of European locations as well as in the Near East in Israel, Lebanon and Jordan. However, as yet, very little scientific research has been carried out on the artefacts found in Egypt (Serpico 2000, 464) and the sources remain unclear.

At the moment, the earliest evidence for any regular use of bitumen by the ancient Egyptians dates to the 18th Dynasty (Serpico and White 2001). One or two sources have now been located in Egypt (Harrell and Lewan 2002) but the current information suggests that they were not exploited until after the New Kingdom. Best known are the sources in the Near East, particularly around the Dead Sea, where bitumen can be obtained in a semi-fluid state from seeps, and solid pieces and blocks have sometimes appeared in the Dead Sea after seismic activity. However, sources from more distant locations such as Iraq, appear to have reached Egypt in later times (Connan and Dessort 1989; 1991).

Of the different classes of natural product, oil and fat sources seem to show the greatest changes in the New Kingdom. Oil sources include a number of fruiting trees and oilseed crops. Of these, it has been suggested that olives may have begun to have been cultivated in Egypt in this period (Hepper 1991, 16; Meeks 1993, 4–5; Keimer 1924, 29–30) but there is still debate regarding the possible introduction of sesame into the Near East and Egypt during the Late Bronze Age (Zohary and Hopf 2000, 140–1; Keimer 1924, 19; Serpico and White 2000b, 397). There is also some evidence to suggest that almond, poppy and safflower were known at this time (Serpico and White 2000b, 393–4, 401, 404), but the extent of cultivation is again unclear. Linseed was used from very early times, and sources such as castor, balanos, and moringa grew wild although there is little indication that they were systematically cultivated.

A variety of animals, both wild and domesticated could also have been exploited for fat (Serpico and White 2000a, 407–8). Notably, it has been suggested that the fat-tailed woolly sheep had become dominant in Egypt during the New Kingdom (Zeuner 1963, 183; Ryder 1983, 109). Another possible introduction, based on pictorial evidence, is the zebu, a type of cow with a distinctive lump of fat behind the neck (Houlihan 1996, 11; Epstein and Mason 1984, 12; Ikram 1995, 13). Conversely, the aurochs seems to have become extinct in Egypt at some point after the reign of Amenhotep III (Ikram 1995, 13).

Wax and honey would have been provided by the honeybee, *Apis mellifera lamarckii*. There is a long history of beekeeping in ancient Egypt with tomb scenes showing manufactured beehives from the 5th Dynasty onward (Crane 1983, 35–39; Crane and Graham 1985, 2–5; Serpico and White 2000b, 409–10). Apart from these scenes, there is little detail on beekeeping practices during the New Kingdom. As bees need to be near a water source, it is understandable that honeycombs and jars of honey are shown in some tombs of this period as products of the Delta (Davies 1930, 34, Pl. XXXI; Davies 1943, Pl. XXXIII–V). Texts indicate that temples also employed beekeepers (Peet 1930, 132 (18)) and the title 'Overseer of the Beekeepers' is also known (Pendlebury 1951, 179). Conversely, it is also clear that, at least during the New Kingdom, honey was imported in quantity from the Near East (*e.g.*, Sethe 1907, 670 [8], 688 [8]; Serpico, 1996, 268–270).

This does not necessarily mean that Egyptian sources were not available. It may be that foreign honey was viewed as more prestigious or tasted differently. In support of this, some jar inscriptions from Amarna suggest that two different types of honey may have been recognized, but the reasons for the distinctions are unclear (Pendlebury 1951, 175).

In summary, private individuals would potentially have had access to many natural products. The materials could be harvested wild or derive from personal small-scale plantings. However, management of sources could have allowed a more regulated supply, with clear economic advantages. This undoubtedly provided an incentive for pharaonic control of these resources. For imported products such as resins, bitumen and some oils, the pharaoh could have exercised a high level of control. For example, without the largess of the king, the average person might be able to obtain oil, but the choices would be limited.

Procurement

Turning now to this stage of production, as mentioned above, substances such as resins and gums can be obtained directly from the source by simple collection of spontaneous exudates or by tapping. This is done by making incisions with a sharp implement and collecting the resin either while it is still in a viscous state or collecting the dried tears. Confirmation of these procedures is absent until much later when it is described by classical writers such as Theophrastus (*Enquiry into Plants*, IX.1–6, *passim*; see Hort 1977) and Pliny (*NH*, XII.32.58–37.77; see Rackham 1968). Modern ethnographic studies have identified a number of specific tools for the tapping, but none are known from antiquity. Moreover, it is important to remember that resins and gum-resins were almost exclusively imported and therefore it would be necessary to look at other Mediterranean sites for material remains.

For other products, a two-step process is necessary: first the combs must be extracted from the hive and then, from them, wax and honey. Methods of procuring oil sources would depend on whether plant seeds or fruits or kernels were used. Similarly, animals must first be butchered before the fat can be rendered. Documentation for the procurement is best for domesticated sources which were exploited for other purposes. For example, because of the importance of meat and its ritual connotations, scenes of butchery often appear in tombs. Scenes depicting the harvesting of linseed do occur (Tylor and Griffith 1894, Pl. III), but probably due to the importance of flax fibre rather than the oil.

With regard to honey, the 18th Dynasty tomb of Rekhmire shows a good example of harvesting honeycombs from man-made cylindrical hives (Davies 1943, Pl. XLIX). Also shown in the scene is a man holding a bowl with burning wicks. This most likely represents smoking the hive to drive the bees out to enable the collection. Ethnographic evidence from earlier this century indicates that hives were harvested twice a year in the spring and late autumn in Egypt but the frequency in ancient times is unknown (Kuény 1950, 90; Mellor 1928, 26).

Processing

Honey can be most easily extracted by placing the combs in bags, crushing them and then letting the honey run out (Mellor 1928, 28–31). The drained combs can then be melted to produce the wax. This can be done over a direct heat, but it better done in water where the wax can be skimmed off as it hardens.

Oil processing would depend of the part of plant producing the oil. For some sources, the fruit would be picked from the trees, and the flesh stripped and the nutlet cracked to obtain oil from the kernel. Some harvested oil seed crops would require threshing, but castor and, if available, sesame, are dehiscent and would spontaneously scatter seeds thus requiring harvesting before maturation and special efforts to retain the seeds. Once collected, oil seeds would need to be cleaned of impurities, and possibly dehulled of their seed coat. This would be followed by grinding or mashing the seeds with a mortar and pestle. Most oil seeds and kernels could be kept for longer periods, thus extending the available period for processing. However, as olive oil is extracted from the fruit, olives need to be pressed soon after harvesting which would intensify activity during this period.

The actual oil extraction can be carried out in a number of ways (Serpico and White 2000b, 406–7). The simplest is to boil the seeds in water and collect the oil as it rises to the surface. However, the best known method is by a form of bag pressing where the ground pulp is placed in a cloth sack and then hung to drain or twisted at both ends to force out the oil into a receptacle below. After the initial pressing, the pulp could be heated and wrung out again. Tomb reliefs suggest that this type of pressing was in use by the Old Kingdom (Lepsius 1849–59, II, Bl. 49), certainly for wine and perfume production. By the Middle Kingdom (Newberry 1894, pl. VI), a more elaborate press was in use, using a wooden framework to increase torsion. It is often assumed that it was used for oil manufacture as well, but thus far, convincing evidence is lacking, as there are no definite examples where oil manufacture is specifically mentioned. For example, a slightly different, sturdier version is attested in a wine pressing scene from the New Kingdom tomb of Puyemre (Davies 1922, Pls. 12–13) but its use in oil manufacture is unknown.

Notably, in the New Kingdom, the title most often used in conjunction with oil manufacture is *ps sgnn*, translated as 'oil boiler,' which would seem to confirm the use of the boiling technique. The title appears quite frequently in the 18th Dynasty, supplanting at least in part the earlier term *nwdw*. This latter term was in use in Middle Kingdom where it not only occurs in titulary, but *nwd* is also the name of an ointment as well as the name of the bag press (Erman and Grapow, 1928, 226; Serpico and White 2000b, 462). This emphasizes the difficulties in determining if and in what circumstances bag pressing was used.

The title *ps sgnn* occurs in hieratic jar labels from Amarna, particularly on imported Canaanite amphorae used to transport *nḥḥ* oil (Petrie 1894, pl. XXIII; Pendlebury 1951, 179; Serpico 1996, 253–4). The jar inscriptions suggest that one duty of the *ps sgnn* was to take delivery of the imported oils for the temple and palace stores, but the title itself implies that the *ps sgnn* was also responsible for the manufacture of the oils.

The botanical identity of *nḥḥ* oil is problematic and both olive and sesame have been proposed (Koura 1995; Krauss 1999). Olive was widely available in the Near East from Chalcolithic times and some have taken the appearance of the word *nḥḥ* at this time as an indication of the introduction date of sesame (Keimer 1924, 19; Manniche 1999, 31). As part of a large project to study Canaanite amphorae imported into Egypt, sherds from jars labeled *nḥḥ* were analysed by Benjamin Stern and Carl Heron, using GC/MS. The results attest to the presence of oil although unfortunately the botanical identity could not be established (Stern *et al.* 2000).

Petrographic study of the *nḥḥ* sherds and other uninscribed examples made from the same clay fabrics has been carried out by Laurence Smith and Yuval Goren, also as part of the Canaanite Amphorae Project (Smith *et al.* 2000; Bourriau *et al.* 2001, 127–37, 142–4). The results have confirmed that some jars inscribed for *nḥḥ* originate in Syria, probably from the environs of Ras Shamra, while others come from coastal Lebanon, indicating that more than one location was active in production at that time.

Current evidence suggests that different methods of pressing to those used in Egypt were used in the Near East (Epstein 1993, 137; Eitam 1993a, 77; Frankel 1994, 28–31). From Chalcolithic times onward the procedure for olive oil manufacture was to place the crushed olives in a basket which was then placed onto a flat-bottomed vat with a deep central depression. A heavy stone weight was positioned on top of the basket to force the oil into the central collecting reservoir. This technique was used in Palestine through the Late Bronze Age, although there is now archaeological evidence that a new important innovation, the lever beam press came into use in Ras Shamra during the Late Bronze Age (Callot 1987, 204–8; 1994, 191–6). This method was similar to the traditional one, except that additional force was supplied by a long wooden beam, inserted into a niche on an adjoining wall just above the baskets (Figure 6.5). To increase the pressure still further, stone weights were hung from the beam. By the Iron Age, this method had become more common, as excavations at the 7th century B.C. site of Tell Miqne Ekron in Israel have shown. There, it has been estimated that at least 230 tons of olive oil could be produced annually (Eitam 1993b, 96; Gitin

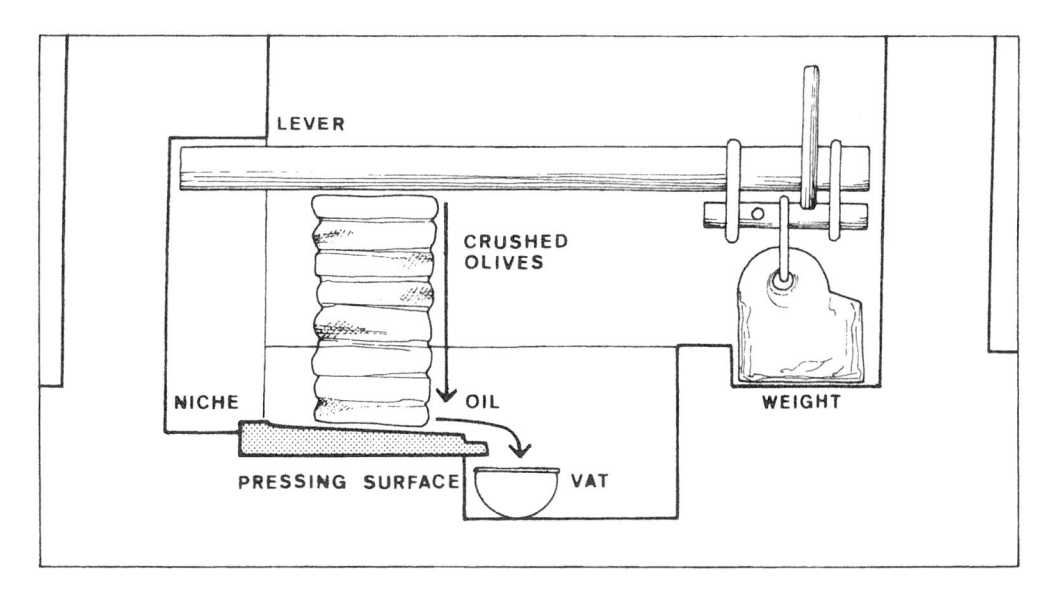

Figure 6.5 Lever press from Ras Shamra, after Callot 1987, 208, Fig. 10. Drawing by Will Schenck.

1990, 40). Large scale production at Ras Shamra would also seem possible based on discovery of a large collection of amphorae in a storage depot at nearby Minet el-Beidha (Schaeffer 1949, Pl. XXXI.1). Unfortunately, only one jar is presently available for study, now in the Louvre, but visual examination has established that it matches the fabric of Syrian origin used to transport *nḥḥ* oil at Amarna (Serpico *et al.* 2003, 372).

Clearly, the effect of this development was potentially quite dramatic. Regardless of whether *nḥḥ* was sesame or olive, it is still evident that the inhabitants of Ras Shamra had developed a technology which could significantly increase production. Can it therefore be a coincidence that *nḥḥ* oil appears in Egypt for the first time concomitantly, and that one of the most common Canaanite amphorae fabrics found at the Workmen's Village at Amarna is that from Ras Shamra (Serpico *et al.* 2003, 374)? Its prevalence attests to the ready availability of this oil outside the royal enclave. References to *nḥḥ* oil become increasingly common thereafter and it was often mentioned as a ration for workers, and frequently exchanged as a commodity between them (Janssen 1975, 330–1). Notably, its description as a standard illuminant also suggests its widespread accessibility.

By the Ramesside period, another change appears to have taken place. At Deir el-Medina, inscribed storage jars allude to *nḥḥ* produced in Egypt, again implying that this oil was a recent Late Bronze Age introduction to Egypt. A number of jars carry the inscription, '*nḥḥ* of the great orchard of *ddw*' produced for the temple (Koenig 1979, Pl. 1–11). Interestingly, the word used here for orchard is *k3mw*, the same word used in conjunction with wine production and translated in that context as vineyard. *Ddw* is usually translated as 'olives,' and the connection in this context to the word for orchard would also suggest that *nḥḥ* is more likely to be associated with olive trees rather than a crop plant like sesame (Krauss 1999). However, it should be noted that the relationship of imported *nḥḥ* to local *nḥḥ* is still unclear. It is also not known whether the new pressing technology moved to Egypt with the botanical source, nor is there any indication in the archaeological record of increased processing of oil sources during this time. Archaeological evidence for any type of press, whether for oil or wine production, is lacking. Although at first surprising, it is also true that many of the implements used are not exclusive to oil production. It would be difficult to attribute wooden poles, baskets, cloth bags, vats, mortars and pestles found on a site to such production and as yet, significant botanical debris which could suggest production centres has also been missing. It is possible that excavators in Near Eastern locations are much more sensitive to identifying remains because of the long history of oil production there.

Also interesting is an association between the *ps sgnn* and fat production. One Amarna jar docket mentions fat made by the *ps sgnn* (Pendlebury 1951, Pl. XCIV.258). This connection is not entirely surprising as fat could have been rendered in a similar fashion to wax and oil. After removal of any extraneous veining and membranes from the fatty tissue, the raw fat could be heated dry or in water and once liquified, mixed with water and left to stand until it separated. Notably, edible animal fats are usually rendered within a few hours of killing to avoid degradation. This suggests that butchery and rendering facilities would have been located in close proximity. However, for most natural products, it is difficult to know how close procurement and processing areas were to each other.

Refinement

After processing, some products, particularly oils, fats, and honey may have been refined. If necessary or desired, any remaining solid debris or contaminants can be removed by hand or by processes such as washing or filtration. This stage of refinement may have been common for natural products, although some cultures prefer unrefined products and today unfiltered oils are sometimes available. Refinement might have been carried out in the processing area, or it might be left to the recipient of the natural product to undertake this.

With regard to the *nḥḥ* oil imported to Egypt in Canaanite amphorae, a number of the hieratic jar labels also relate the *ps sgnn* to an action designated by the verb *sw3b*. The meaning of *sw3b* here is somewhat unclear. Specifically, it most often means 'to cleanse or purify,' which could allude to some stage of refinement that was carried out by the *ps sgnn*. If so, this would raise questions of whether the imported oil needed refinement, and when and where the refinement was carried out. With regard to Canaanite amphorae, it is often thought the inscriptions were put on the jars when they first arrived in Egypt and not at their final destination, although documentation before initial shipping has also been argued (Bavay *et al.*

2000, 80–1). This also raises the possible logistical issue of re-decanting the oils back to the Canaanite amphorae after refining. However, *sw3b* may also mean 'to consecrate,' suggesting a possible religious procedure, perhaps necessary before accepting the oils into the temple or palace specified in the inscription. If so, this would indicate an overlap between secular and religious functions.

Interestingly, an inscription on a Canaanite amphora found at Amarna (Frankfort and Pendlebury 1933, Pl. LVIII, 36; Serpico 1996, 104, Fig. 6–1) gives the contents as incense, *sntr*, and also mentions *sw3b* and the *ps sgnn*. Imported resin might have required some refinement to remove trapped debris. Study of the Canaanite amphorae at Amarna with visible resin residues revealed no obvious embedded contaminants, but a number of small filament shaped pieces have been found at the site (Figure 6.6). These appear to have been passed through a sieve and indeed they have diameters that match the average size of sieve holes (Serpico 1996, 144–6). Chemical analysis of one of these samples, along with resin

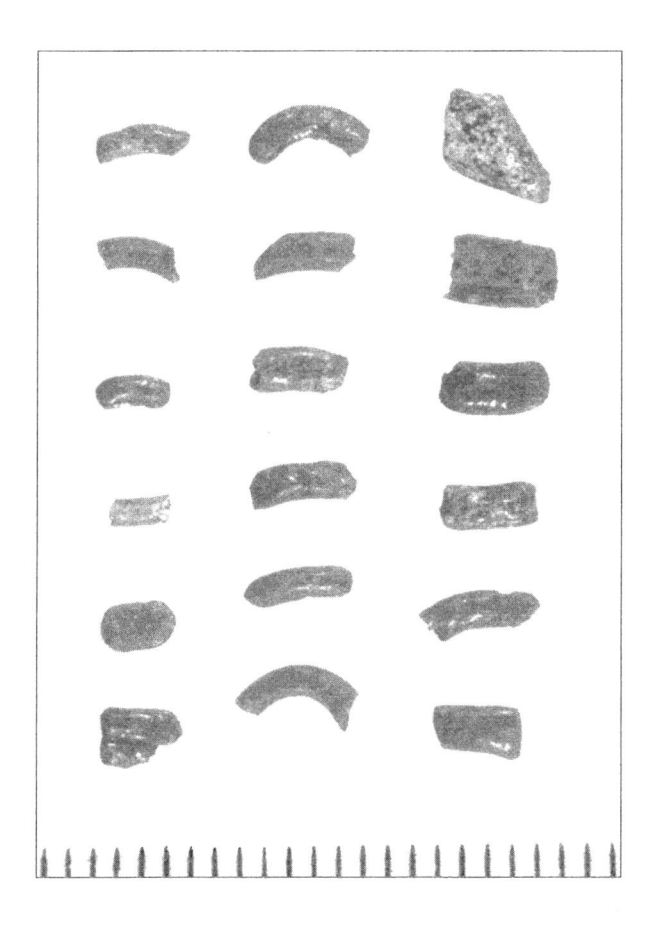

Figure 6.6 Deposit of resin fragments from Amarna. Measurement at the bottom in mm.

deposits in incense bowls and residues coating Canaanite amphorae labelled *sntr* has confirmed that the source was a species of *Pistacia*, possibly *Pistacia atlantica* (Serpico 1996; Serpico and White 2001a). Moreover, the clay fabrics used to manufacture these imported jars differ from those used to transport *nhh* oil. Two fabric groups are associated with the transport of *sntr*, both located in what is today northern Israel (Smith *et al.* 2000; Bourriau *et al.* 2001, 140, 143; Serpico *et al.* 2003, 368–69). One of these fabrics closely resembles that used for the Canaanite amphorae found on the Uluburun shipwreck which also held pistacia resin (Serpico and White 2000a; Mills and White 1989). Notably, many of the shipwreck jars had evidence of debris mixed with the resin (Bass 1986, 278). Given that collectively over a ton of pistacia resin was carried on the Uluburun shipwreck (Pulak 1998, 201) and given that *sntr* was a standard component in Egyptian daily temple rituals, it is not difficult to imagine that the scale of importation was great (Serpico 2003). The presence of dozens of resin-coated sherds and resin pieces at the Workmen's Village at Amarna again attests to the availability of this product outside the royal circle. For the moment, however, filament shaped pieces appear to be more common at the Small Aten Temple (Serpico 1996, 173–213).

The interpretation of the jar labels is further complicated by changes in the personnel involved. Some Canaanite jars inscribed for incense mention *sw3b* followed by a reference to a guard (*s3w*) rather than the *ps sgnn* (Frankfort and Pendlebury 1933, Pl. LVIII, 34; Pendlebury 1951, Pl. XCVI, 288, 289; Serpico 1996, 704–6) and one example associates *sw3b* with a *ps š'yt* (Frankfort and Pendlebury 1933, Pl. LVIII, 35) a relatively common title translated as 'baker of cakes.' It is evident that there is much not understood about the process described as *sw3b* but the connection of the *ps sgnn* to oil, fat and resin deserves further study.

In light of this, it is possible to see the *ps sgnn* as an individual attached to a temple and/or palace workshop involved in the production of oils and of ritual mixtures. In fact, *sgnn* can refer not only to a simple oil, but also to a scented ointment (Erman and Grapow 1930, 322; Janssen 1975, 336–7; Helck 1963, 504). In this, the word seems to mirror the word *mrht*, which was used from Old Kingdom times to refer both to oil in general and to ritually intended complex unguents. This again emphasizes the potential for confusing the circumstances in which the bag press was used.

Packaging

Once formed, a simple natural product may be packaged for distribution or transferred to another processing area. A variety of storage methods would have been available depending on whether the final product was fluid or dry. Most natural products would be in a fluid or semi-fluid state at this stage and storage in pottery jars would have been most suitable. Although other types of vessel, such as stone and glass, would have been available, the use of these is less likely for bulk transport of simple natural products. Others substances, such as resin, wax, or bitumen, may have been left to dry or have been collected as hard pieces initially. In this case, storage in bags, baskets, skins or other containers would perhaps have been more practical.

COMPLEX NATURAL PRODUCTS

While simple natural products would have been widely used, complex mixtures to form cosmetics, ritual unguents and so on (Figures 6.2, 6.4) were also important. In the case of resins, wax or fat, the simple products could also be manipulated into recognizable forms.

Thermal processing

For some products, a stage of thermal processing would have been needed, for example to soften or liquefy substances to facilitate manipulation or admixing, or to form resin pitches by high temperature heating. This was most likely done in a vessel over a fire. Pitch, of course, could then have been packaged and distributed on its own or again shaped into forms, as discussed below.

Admixing

In the preparation of religious or ritual ointments, medicinal treatments or cosmetics, the range of possible added components is virtually endless and could have included mixtures of different natural products, as well as other organic compounds such as scented flowers and herbs or even inorganic components such as ground minerals. That such a variety of components was used is attested in later texts, notably from the 'laboratories' at Edfu and Dendara (Chassinat 1990; Dümichen 1879). These could have been mixed directly into a simple natural product base to create a complex natural product, or the mixture could then undergo processing treatment.

Processing

Further processing, including re-heating, pressing or refinement could then follow. Little is known about such procedures during the New Kingdom although, again, later Ptolemaic texts confirm the complexity of the manufacture of these substances (Dümichen 1897; Chassinat 1990; Baum 2003, 78–82; Aufrère 2003, 144–151). These processes might also have fallen under the jurisdiction of individuals such as the *ps sgnn*. In addition to bag pressing as discussed above, the techniques of enfleurage or maceration may also have been utilized to produce scented ointments (Dayagi-Mendels 1989, 97, 100; Lucas 1962, 86; Serpico 2000, 461). In enfleurage, animal fat is applied to a wooden board and flowers or petals are placed across the surface. These are then pressed with another board and left for the scent to infuse the fat. The flowers or scented material would be changed regularly and the process could continue for a number of weeks. Maceration is a technique of hot steeping, similar to a modern double boiler. The aromatic substances are mixed in a vessel with fat or oil, and this is set above a heated pan of water. Over time, the mixture would be stirred and reheated. This process was described by Theophrastus (*Odours*, 22; Hort 1977, 347). Some components, such as resins and gum-resins, might dissolve in the heated lipid matter, but other mixtures might then require pressing or refinement.

Refinement

As in the manufacture of some simple natural products, it might be necessary in some instances to remove unwanted solid debris by hand-picking, straining or by letting contaminants settle to the bottom of a container and carefully pouring off the bulk of the mixture. This process might have been carried out a number of times for complex mixtures.

Scenes depicting stages in the manufacture of scented ointments are comparatively rare (Lucas 1962, 86). For the New Kingdom, a scene from Theban tomb 175, dated to the time of Thutmosis IV, is believed to show the production of a scented cosmetic (Manniche 1989, 56–7; Baum, 2003, 78, Fig. 5). Unfortunately, there are no inscriptions accompanying the scenes, but Manniche (1989, 57) suggests that the sequence shows a series of processes including: scraping chips of aromatic wood, letting them steep in wine and then straining them; melting fat in a vessel, mixing aromatics with the fat, placing this mixture in a pot and boiling it gently. However, while such a scenario is plausible, there is no means of determining whether any of the actual processes are condensed or omitted here, or whether the activities are depicted in the correct order. In fact, for the period of the New Kingdom, we do not know the extent to which any of the processes such as admixing, heating and straining might have been repeated in the preparation of a single scented ointment, nor do we know whether the sequence of these procedures varied.

The increasing number of analyses of jar contents and of residues found on mummies also demonstrates the compositional variation of these scented unguents (recently, Tchapla 2003, 152–61, also a summary in Serpico 2000, 456–467). Samples taken from mummies, almost exclusively of later date, indicate that a number of products, such as coniferous resins and pitch, lipid substances, beeswax and pistacia resin were applied. Only one possible sample of bitumen from a New Kingdom mummy is known (Connan and Dessort 1991, 1449), but the source of that sample is not well documented (Serpico 2000, 466). Variation in the samples taken from different parts of mummies suggests that different simple and/or complex natural products were applied to different areas of the body, but as yet no clear patterns connected with specific areas of the body have emerged. Similarly, analysis of jar contents can be problematic as it is often impossible to determine the intended use of the substance. Jar contents could relate to religious, medicinal or cosmetic usage. While random analyses of jar contents can provide information on the range of products in exploited in a certain period, they cannot necessarily shed light on their intended function.

One use of natural products which has been more systematically explored is the application of these substances on funerary equipment. In the past, these coatings have been described as varnish. Two 'varnishes' were used, a black varnish and a yellow varnish which might originally have been of a lighter colour but has now darkened. A range of objects were coated including coffins, shabtis and shabti boxes, canopic chests, stelae, statues, painted vases and, in the case of yellow varnish, even some tomb walls (Lucas 1962, 356–61). This practice continued through the New Kingdom, becoming less prominent towards the end of the period, but resuming with force in the Third Intermediate Period. It was so widespread that today it is hard to find a museum collection that does not have at least one varnished object, indicating that these substances were available to a range of private individuals.

To determine the identity of the varnishes, samples were taken from a range of New Kingdom objects and analysed by Raymond White using GC/MS (Serpico and White 2001, 34–6). The yellow varnish consisted most often of only pistacia resin, although in one instance this had been mixed with lipid matter. Textual evidence suggests that the yellow pistacia resin varnish was also called *sntr*, the same word that is usually translated as incense, but from these results it is clear that it is better to take the word literally, as 'to cause to make divine.' In that sense, *sntr* is not just incense but a substance intended to confer divinity on offerings to the gods and the deceased.

The black varnish showed considerably more variation. While the coating on some objects consisted of a strongly heated pistacia resin, which could probably be termed a pistacia pitch, other objects contained mixtures including components such as coniferous resin and pitch, pistacia resin and pitch and lipid matter. Two 18th Dynasty samples contained evidence of bitumen. One, the coffin of Tamyt (British Museum, EA 6661), and the second, a canopic box (British Museum, EA 35808) where the black varnish was identified as pistacia pitch and bitumen.

The ancient Egyptian word *mnn* has been translated as bitumen (Loret 1894, 157–162; Aufrère 1991, 639). It first appears in the New Kingdom, and a relief in the tomb of Rekhmire, contemporary with Thutmosis III, depicts lumps of a black substance, heaped in a basket, above which is inscribed the word *mny*, possibly a variant writing (Davies 1943, Pl. XXI). There is also evidence to suggest that the black varnish was called ꜥ3t *ntr* or the 'divine stone' and that its recipe could vary, but did at least sometimes include *mnn* (Chassinat 1955, 65–74; Aufrère 1991, 329–347; Serpico and White 2001, 36–7). As yet, it is still not possible to assess changes in the composition of these substances over time as sample numbers are quite small. However, given the degree of compositional variation already apparent, it may well be the case that no overall consistency will be detected. This is in itself informative as it implies that there was not a set recipe for this mixture, and that perhaps personal choice or availability could influence the composition.

From this study of consecrated funerary objects, it is clear that a range of natural products was used, from simple natural products to a variety of complex natural products. Methods for producing these substances undoubtedly varied, although in some cases they were serving the same purpose. For example, it is evident that the coating of *Pistacia* sp. resin as 'varnish' overlaps with its usage as incense. As research continues, it will also be interesting to determine whether the complex black anointing mixtures overlap with substances used in mummification. This raises the question of the extent to which the individuals manufacturing these mixtures appreciated the possible multi-purpose nature of their preparations.

Manipulation

The manipulation of natural products into shapes, by hand shaping, moulding or carving, was fairly common in ancient Egypt. Although this could be viewed as another stage in processing, it could also have been carried out separately. Wax figures, with purported magical powers and a long history of use in Egypt (Raven 1983) may have been made subsequent to distribution of wax blocks, but others, particularly those with funerary connections, may have been produced in workshops. Because of its flammable nature, wax was especially associated with destructive magic and curses and perhaps as a result of this, few archaeological remains survive. Amber is another substance which could be carved, particularly into beads and amulets, although it is difficult to distinguish amber from resin by eye and further scientific analyses are needed. In the New Kingdom, resin used as incense was shaped not only as balls and cones, but also more imaginatively as cattle, geese or loaves of bread (Helck 1963, 513–517). Like wax objects, these were intended to be burned and hence there are no preserved examples in the archaeological assemblage. Other objects said to have been made of resin include heart amulets and beads (Raven 1990).

During the 20th Dynasty, a title, initially transcribed as *ps sntr*, first appears. Translation of this title has been problematic. Although often read as 'incense burner,' Quaegebeur suggested that the first sign should not be read as *ps* but as *s3k*, and suggested that the title relates not to the heating or burning of resin but to the modeling of incense into designated shapes (Quaegebaur 1993, 29–44). If so, the *s3k sntr* would be 'the modeler of *sntr*.' The hierarchical relationship of the *s3k sntr* to the *ps sgnn* is uncertain, but both seem to have been attached to temples. In papyrus BM 10068, which gives the town register for the area around Medinet Habu, two people are listed with the title *s3k sntr* (as well as three people with the title *ps sgnn*, and three listed as beekeepers) but it is not known how many people could hold this title in one temple at a given time (Peet 1930, 95–98; see also Shaw, this volume, for a more detailed discussion).

Unfortunately, the place where we hear the most about these individuals is in the New Kingdom Tomb Robbery Papyri. In BM 10052, three people with the title *s3k sntr* were brought up on charges of robbery, one of whom, associated with the temple of Amun, appears again in Abbott Docket A.23 (Peet 1930, 132, 143–149). Also mentioned in the trial are an oil boiler and a beekeeper, and it is clear that others with these titles were also involved in tomb robberies or held stolen goods (Peet 1930, 91, 92, 95, 97, 98, 106, 132, 171).

Although partially damaged, one tomb scene appears to recount the process of making these modeled shapes. In the tomb of Amenmose (TT89) at Gourna, dated to the reign of Amenhotep III, a series of scenes depict what has been described as "the making of incense, salves, pomades, &c., for a very large household" (Davies and Davies 1940, 133; Baum 2003, 70). Amenmose held none of the known titles associated with natural products, but Davies and Davies (1940, 132) noted that the "kitchen scene is not inappropriate to the tomb of a steward who was responsible for the domestic economy of the palace." There is one possible link to these scenes: recent re-examination of the tomb suggests that some of the reliefs were possibly covered with a coat of the yellow 'varnish' discussed previously (Brock and Shaw 1997, 169). While no scientific examination of this coating has been undertaken, the link between the 'varnish,' *Pistacia* sp. resin, and these modeled shapes is possible.

The top register in the tomb (Figure 6.7) shows the following (Davies and Davies 1940, 133): 1) a scribe records a delivery of goods (only traces survive), 2) two men pour ingredients into a cooking pan, 3) the pan is carried off, 4) the pan is placed on top of a stove and stirred by one man, while another pours ingredients from a jar, 5) continued stirring of the contents or perhaps ladling out of them while a child stands by, 6) shaping into a mass on a small table, 7) moulding into the shape of a trussed fowl, 8) depiction of a man who is perhaps overseeing the process. The next two registers show men carrying an array of moulded shapes including cones, oxen, and trussed fowl. Another register (Figure 6.8) then seems to demonstrate show another similar sequence of processing: 1) a scribe writes a record of storage jars, 2) while standing over a pan, a man pours the contents of a jar into a strainer held by another man, 3) the pan is carried by two men, 4) the pan is placed on a stove and heated while a man stirs the contents and, as suggested by the presence of a vessel nearby, possibly adds to the mixture, 5) an overseer presides over the continued stirring of the contents, 6) an array of moulded forms including oxen, gazelles, trussed fowl and obelisks is displayed. Interestingly, a large figure of what may be a zebu, mentioned above as possibly having been introduced in the New Kingdom, is shown.

While these scenes appear to be quite detailed in one respect, just as for those in TT175, it is important to remember that stages of production are almost certainly omitted, and the sequence may also be incorrect. It is interesting that

Figure 6.7 Scenes from the top register of Wall F in the tomb of Amenmose, after Davies and Davies 1940, pl. XXII. Drawing not to scale.

many of the same vignettes of heating and stirring occur in both. In the case of the reliefs from Amenmose, the absence of any helpful text is especially unfortunate, as clues to the titles held by the individuals involved in the processing would go some way to resolving the current problems in understanding this industry and the confusion between *ps snṯr* and *s3k snṯr*. Nonetheless, we can see that the process involved at least one scribe, two people carrying out the actual production and one overseer. If this does indeed represent the transformation of *snṯr*, or *Pistacia* sp. resin into moulded shapes, then the process would seem to represent the decanting of fluid resin into a strainer to remove any debris, the placing of a pan of the resin onto a stove, the heating of the resin to the necessary consistency and the moulding of the resin into forms. It seems possible that other ingredients were added, but as no actual examples of these forms still exist, there is no means of determining the composition of the finished products.

Packaging

Again, a processed mixture could then have been packaged and distributed or once in a fluid or malleable state, manipulated into shapes. As well as pottery vessels, baskets, bags, etc., in these instances, more prestigious containers such as stone vases and glass vessels might have been used.

Figure 6.8 Scenes from the fourth register of Wall F in the tomb of Amenmose, after Davies and Davies 1940, pl. XXII. Drawing not to scale.

DISCUSSION

In conclusion, there were clearly a number of innovations in the processing and use of natural products during the New Kingdom, with the introduction of new natural product sources, possible indications of foreign influence on production, technological advances, new compositions and uses, and even new titulary. But the scope for further research is noteworthy. Particular areas for advancement include more data on the individuals involved in the processes, more lexicographical information on the products and greater consideration of the role of storage and transport in the overall technological process. In addition, given the influence of foreign contact, a comparison of the personnel and procedures used in other Mediterranean regions might also prove instructive.

In this regard, it is particularly interesting to explore within a broader context the introduction of the use of 'varnish' on funerary equipment. This new development in the use of natural products, dated towards the end of Hatshepsut's reign, begins at a time when the assemblage of funerary equipment is also changing stylistically (Taylor 2001, 168-69; Serpico 1996, 324–331). It is, however, not clear whether the changes in funerary equipment gave rise to the use of the varnish, whether the introduction of abundant supplies from abroad suddenly made such widespread usage more feasible, or a combination of the two. Conversely, this new usage would have greatly increased the need to import these goods. The *Annals* of the campaigns of Thutmosis III, which list substantial quantities of imported oil, resin and honey, intimate that he was willing to ensure that these supplies did reach Egypt (*e.g.*, Sethe 1907, 688 (8); 693 (1); 694 (5); 699 (15); Serpico 2003, 228). Other important changes were occurring at this time as well. It is during his reign that glass first began to be imported in quantity, with the corresponding initial appearance of a glass industry in Egypt (Nicholson and Henderson 2000, 195; Shortland *et al.* 2001, 151; Shortland 2001, 211). Glass, of course, would have been one of the most elegant containers of aromatic preparations. Stylistic changes and new forms, possibly introduced from the Near East, also appear in the pottery and stone vessel repertoire, particularly those shapes associated with the storage of aromatic ointments (Spalinger 1982, 126; Bourriau 1982, 127–9; 1981, 72). Thutmosis III's campaigns may well have influenced or intensified the development of a number of related industries. For instance, scenes of chariot manufacture first appear in tombs dated to the reign of Hatshepsut and continue under his reign (Shaw 2001, 63). These instances raise the possibility that the time from Hatshepsut to Thutmosis III marks the transition in some technologies from introduction and limited production to adoption and a more structured industrial approach.

Conversely, a decrease in foreign contact is often suggested during the reign of Akhenaten, but the quantities of natural products reaching Amarna in imported Canaanite amphorae indicate that supplies were clearly available. During that time, Egypt probably retained influence in areas producing *Pistacia* sp. resin, but connections to areas supplying *nḥḥ* oil would have been less reliable (Serpico and White 2000a, 895–896). Unfortunately, at the moment and in the absence of comparable data from an earlier date, it is difficult to know how to interpret the chronological significance of this data. We simply cannot quantify the amounts shipped during the Amarna period or the preceding periods with sufficient accuracy to determine changes in quantities supplied over time. But it is nonetheless clear that supplies did reach the Workmen's Village in quantifiable amounts.

In the case of natural products, scientific analyses of preserved residues and study of floral and faunal remains can provide new insight into natural products and their uses. At the moment, our understanding of the technologies used is somewhat limited as, in some cases, we are not certain which sources were exploited and processing techniques can vary. By clarifying the identity of the substances and the contexts in which they were used, we can begin to explore the technologies which produced them in greater detail. This in turn can help us understand the role played not only by these commodities, but also by the methods of manufacture, in ancient Egyptian society.

ACKNOWLEDGMENTS

I would like to thank the Egypt Exploration Society, Mr. Barry Kemp and Dr. Pamela Rose (both McDonald Institute of Archaeological Research, Cambridge) for access and help with the Amarna material. For permission to export material for study, I would sincerely thank the Supreme Council for Antiquities, Egypt. Collaborators on the Canaanite Amphorae Project are Ms. Janine Bourriau and Dr. Laurence Smith (both McDonald Institute of Archaeological Research, Cambridge), Dr. Carl Heron and Dr. Benjamin Stern (Department of Archaeological Science, University of Bradford) and Dr. Yuval Goren (Depart-ment of Archaeology, Tel Aviv University). This project has been funded by the Wainwright Fund of the Oriental Institute at the University of Oxford, the NERC, the Egypt Exploration Society, the McDonald Institute for Archaeological Research, the British Academy and the Society of Antiquaries. I would like to thank the following museums for permission to study and sample museum objects: the Petrie Museum of Egyptian Archaeology (Dr. Stephen Quirke and the late Mrs. Barbara Adams), the Ashmolean Museum (Dr. Helen Whitehouse), the British Museum (Dr. John Taylor, Mr. W. V. Davies and the Scientific Department), Manchester Museum (Dr. Rosalie David) and the Louvre (Dr. A. Coubert). My gratitude also goes to Mr. Raymond White, who has been very generous with his time and expertise.

REFERENCES

Aufrère, S., 1991, *L'univers minéral dans la pensée égyptienne*, I–II. L'Institut français d'Archéologie orientale, Cairo.
Aufrère, S., 2003, Nature et emploi des parfums et onguents liturgiques: recettes, in M.-C. Grasse (ed.), *L'Égypte: Parfums d'histoire*, 144–151. Somogy éditions d'art, Paris.

Bass, George F., 1986, A Bronze Age Shipwreck at Ulu Burun: 1984 Campaign, *American Journal of Archaeology*, 90, 269–96.
Baum, N., 2003, Le Transformation des matières, in M.-C. Grasse (ed.), *L'Égypte: Parfums d'histoire*, 70–83. Somogy éditions d'art, Paris.
Bavay, L., Marchand, S. and Tallet, P., 2000, Les jarres inscrites du Nouvel Empire provenant de Deir al-Médina, *Cahiers de la céramique Égyptienne*, 6, 77–86.
Bourriau, J., 1981, Umm el-Ga'ab: Pottery from the Nile Valley before the Arab Conquest. Fitzwilliam Museum, Cambridge.
Bourriau, J., 1982, Amphora (entries 114, 115, 116), Vase (entry 117), Pitcher (entry 118), Goblet (entry 119) and Dish on stand (entry 120), in *Egypt's Golden Age: The Art of Living in the New Kingdom 1558–1085 B.C.*, 127–129. Museum of Fine Arts, Boston.
Bourriau, J., Smith, L., and Serpico, M., 2001, The provenance of Canaanite amphorae found at Memphis and Amarna in the New Kingdom, in A. J. (ed.), Shortland, *The Social Context of Technological Change: Egypt and the Near East 1650–1150 BC.*, 113–46. Oxbow Books, Oxford.
Brock, L. P. and Shaw, R. L., 1997, The Royal Ontario Museum Epigraphic Project – Theban Tomb 89 Preliminary Report, *Journal of the American Research Center in Egypt*, 34, 1997, 167–178.
Callot, O., 1987, Les huileries du Bronze Récent a Ougarit: premiers éléments pour une étude, in M. Yon (ed.) *Le Centre de la Ville. Ras Shamra-Ougarit*, III, 197–212. Editions Recherche sur les Civilisations, Paris.
Callot, O., 1994, *Le tranchée 'Ville Sud': Etude d'architecture domestique. Ras Shamra-Ougarit X*. Editions Recherche sur les Civilisations, Paris.
Chassinat, É., 1955, *Le manuscrit magique Copte: No 42573 du Musée egyptien du Caire*, L'Institut français d'Archéologie orientale, Cairo.
Chassinat, É., 1990, *Le temple d'Edfou*, II, 2nd edn., rev. Sylvie Cauville and Didier Devauchelle. L'Institut français d'Archéologie orientale, Cairo.
Connan, J. and Dessort, D., 1989, Du bitume de la Mer Morte dans les baumes d'une momie égyptienne: identification par critères moléculaires, *Comptes rendus de l'académie des sciences*, Série II, 309(17), 1665–72.
Connan, J. and Dessort, D., 1991, Du bitume dans des baumes de momies égyptiennes (1295 av. J.-C. – 300 ap. J.-C.): détermination de son origine et évaluation de sa quantité. *Comptes rendus de l'académie des sciences*, Série II, 312(12), 1445–52.
Crane, E., 1983, *The Archaeology of Beekeeping*. Duckworth, London.
Crane, E. and Graham, A. J., 1985, Bee hives of the ancient world. *Bee World*, 66, 25–41, 148–170.
Davies, N. de G., 1922, *The Tomb of Puyemre at Thebes*, I. Metropolitan Museum of Art, New York.
Davies, N. de G., 1930, *The Tomb of Ken-Amun at Thebes*, II. Metropolitan Museum of Art, New York.
Davies, N. de G., 1943, *The Tomb of Rekh-mi-re at Thebes*, II. Metropolitan Museum of Art, New York.
Davies, N. de G. and Davies, N. M., 1940, The Tomb of Amenmose (No. 89) at Thebes, *Journal of Egyptian Archaeology*, 26, 131–136.
Dayagi-Mendels, M., 1989, *Perfumes and Cosmetics in the Ancient World*. The Israel Museum, Jerusalem.
Dixon, D. M., 1969, The Transplantation of Punt Incense Trees in Egypt, *Journal of Egyptian Archaeology*, 55, 55–65.
Dümichen, J., 1879, Ein Salbolrecept aus dem Laboratorium des Edfutempels, *Zeitschrift*

für Ägyptische Sprache und Altertumskunde, 17, 97–128.
Eitam, D., 1993a, 'Between the [olive] rows, oil will be produced, presses will be trod...' (Job 24,11), in M.-C. Amouretti and J.-P. Brun (eds.), *Oil and Wine Production in the Mediterranean Area*, 65–90. Ecole Français d'Athènes, Athens.
Eitam, D., 1993b, Selected oil and wine installations in ancient Israel, in M.-C. Amouretti and J.-P. Brun (eds.), *Oil and Wine Production in the Mediterranean Area*, 91–106. Ecole Français d'Athènes, Athens.
Epstein, C., 1993, Oil production in the Golan Heights during the Chalcolithic period, *Tel Aviv*, 20, 133–146.
Epstein, H. and Mason, I. L., 1984, Cattle, in I. L. Mason (ed.), *Evolution of Domesticated Animals*, 6–27. Longman, London.
Erichsen, W., 1933, *Papyrus Harris I: Hieroglyphische Transkription*, Éditions de la fondation égyptologie reine Élisabeth, Brussels.
Erman, A., and Grapow, H., 1928, *Wörterbuch der Aegyptischen Sprache*, II. J.C. Hinrichs, Leipzig.
Erman, A., and Grapow, H., 1930, *Wörterbuch der Aegyptischen Sprache*, III. J.C. Hinrichs, Leipzig.
Evans, K. and Heron, C., 1993, Glue, Disinfectant and Chewing Gum: Natural Products Chemistry in Archaeology, *Chemistry and Industry*, 12, 446–449.
Frankel, R., 1994, Ancient oil mills and presses in the land of Israel, in E. Ayalon (ed.) *History and Technology of Olive Oil in the Holy Land*, 19-89. Oléarius Editions, Arlington, VA.
Frankfort, H. and Pendlebury, J. D. S., 1933, *The City of Akhenaten*, II. Egypt Exploration Society, London.
Gitin, S., 1990, Ekron of the Philistines, part II: olive-oil suppliers to the world, *Biblical Archaeology Review*, March/April, 33–42, 59.
Grandet, P., 1994, *Le Papyrus Harris*, I, L'Institut français d'Archéologie orientale, Cairo.
Harrell, J. A. and Lewan, M. D., 2002, Sources of Mummy Bitumen in Ancient Egypt and Palestine, *Archaeometry*, 44/2, 285–293.
Hepper, F. N., 1969, Arabian and African Frankincense Trees, *Journal of Egyptian Archaeology*, 55, 66–72.
Helck, W., 1963, *Materialien zur Wirtschaftsgeschichte des Neuen Reiches*, IV. Akademie der Wissenschaften und der Literatur, Mainz.
Hepper, F. N., 1990, *Pharaoh's Flowers: The Botanical Treasures of Tutankhamun*. HMSO, London.
Hort, A. F. (transl. and ed.), 1977, *Theophrastus: Enquiry into Plants*, II. Loeb Classical Library, William Heinemann Ltd., London.
Houlihan, P. F., 1996, *The Animal World of the Pharaohs*. Thames and Hudson, London.
Ikram, S., 1995, *Choice Cuts: Meat Production in Ancient Egypt*. Peeters, Leuven.
Janssen, J. J., 1975, *Commodity Prices from the Ramessid Period*. E. J. Brill, Leiden.
Keimer, L., 1924, *Die Gartenpflanzen im Alten Ägypten*, I. Hoffmann und Campe Verlag, Berlin.
Koenig, Yvan, 1979, *Catalogue des étiquettes de jarres hiératique de Deir el-Médineh: Nos. 6000–6241*, Fasc. I, L'Institut français d'Archéologie orientale, Cairo.
Koura, B., 1995, Ist $b3q$ Moringaöl oder Olivenöl?, *Göttinger Miszellen*, 145, 79–82.
Krauss, R., 1999, $Nh(h)$-Öl = Olivenöl, *Mitteilungen des Deutchen Archäologischen Instituts, Abteilung Kairo*, 55, 293–298.
Kuény, G., 1950, Scènes apicoles dans l'ancienne Egypte, *Journal of NearEastern Studies*, 9, 84–93.

Lepsius, K. R., 1849–59, *Denkmaeler aus Ägypten und Äthiopien*, II. J. C. Hinrichs, Leipzig.
Loret, V., 1894, Études de drogueriè égyptienne, *Recueil de Travaux*, 16, 134–162.
Lucas, A., 1962, *Ancient Egyptian Materials and Industries* (rev. J. R. Harris). Edward Arnold Ltd., London.
Manniche, L., 1989, *An Ancient Egyptian Herbal*. British Museum Press, London.
Manniche, L., 1999, *Sacred Luxuries: Fragrance, Aromatherapy and Cosmetics in Ancient Egypt*. Opus Publishing Ltd., London.
Meeks, D., 1993, Oléiculture et viticulture dans l'Égypte pharaonique, in M.-C. Amouretti and J.-P. Brun (eds., *Oil and Wine Production in the Mediterranean Area*, 3–38. École français d'Athènes, Athens.
Mellor, J. E. M., 1928, Beekeeping in Egypt: Part I. An account of the Beladi Craft, that is to say the Craft native to the Country with observations made upon it from September 1926 to January 1928, *Bulletin de la Société Royale Entomologique d'Égypte*, XII, 17–33.
Mills, J. S., and White, R., 1989, The Identity of the Resins from the Late Bronze Age Shipwreck at Ulu Burun, *Archaeometry*, 31, 37–44.
Murray, M. A., Fruits, Vegetables, Pulses and Condiments, in P. Nicholson and I. Shaw, (eds.), *Ancient Egyptian Materials and Technologies*, 609–655. Cambridge University Press, Cambridge.
Naville, E., 1898, *The Temple of Deir el Bahari*, III, Egypt Exploration Society, London.
Newberry, P. E., 1894, *Beni Hasan*, II, Egypt Exploration Society, London.
Newman, R., and Serpico, M., 2000, Adhesives, in P. T. Nicholson and I. Shaw (eds.), *Ancient Egyptian Materials and Technologies*, 475–494. Cambridge University Press, Cambridge.
Nicholson, P. T and Henderson, J., 2001, Glass, in P. T. Nicholson and I. Shaw (eds.), *Ancient Egyptian Materials and Technologies*, 195–224. Cambridge University Press, Cambridge.
Peet, T. E., 1930, *The Great Tomb-Robberies of the Twentieth Egyptian Dynasty*. Clarendon Press, Oxford.
Pendlebury, J. D. S., 1951, *The City of Akhenaten*, III. Egypt Exploration Society, London.
Petrie, W. M. F., 1894, *Tell el-Amarna*. Methuen and Co., London.
Pulak, †., 1998, The Uluburun shipwreck: an overview, *International Journal of Nautical Archaeology*, 27, 188–224.
Quaegebaur, J., 1993, Conglomérer et modeler l'encens (s3ḳ snṯr), *Chronique d'Égypte*, 68, 29–44.
Rackham, H. (transl. and ed.), 1968, *Pliny the elder: Natural History*. Loeb Classical Library, William Heinemann Ltd., London.
Raven, M. J., 1983, Wax in Egyptian Magic and Symbolism, *Oudheidkundige Mededelingen uit het Rijksmuseum van Oudheden te Leiden* 64, 7–47.
Raven, M. J., 1990, Resin in Egyptian Magic and Symbolism, *Oudheidkundige Mededelingen uit het Rijksmuseum van Oudheden te Leiden*, 70, 7–22.
Ryder, M. L., 1983, *Sheep and Man*. Duckworth, London.
Schaeffer, C. F. A., 1949, *Ugaritica II. Nouvelles études relatives aux découvertes de Ras Shamra*. P. Geuthner, Paris.
Serpico, M., 1996, *Mediterranean Resins in New Kingdom Egypt: A Multidisciplinary Approach to Trade and Usage*, Unpublished thesis, University College London.
Serpico, M., 2003, Quantifying Resin Trade in the Eastern Mediterranean during the Late Bronze Age, in K. P. Foster and R. Laffineur. (eds.), *Metron. Measuring the Aegean Bronze Age. Proceedings of the 9th International Aegean Conference, Yale University 18–21 April, 2002*. Annales d'archéologie égéenne de l'Université de Liège. Université de Liège and

University of Texas at Austin, Liège and Austin, 224–230.

Serpico, M., Bourriau, J., Smith, L., Goren, Y., Stern, B., and Heron, C., 2003, Commodities and Containers: A Project to Study Canaanite Amphorae imported into Egypt during the New Kingdom, in M. Bietak (ed.), *The Synchronisation of Civilisations in the Eastern Mediterranean in the Second Millennium B.C.*, II, 365–375. Österreichischen Akademie der Wissenschaften, Wien.

Serpico, M. and White, R., 2000, Resin, Pitch and Bitumen, in P. T. Nicholson and I. Shaw (eds.), *Ancient Egyptian Materials and Technologies*, 430–474. Cambridge University Press, Cambridge.

Serpico, M. and White, R., 2000a, The Botanical Identity and Transport of Incense during the Egyptian New Kingdom, *Antiquity*, 74(286), 884–97.

Serpico, M. and White, R., 2000b, Oil, Fat and Wax, in P. T. Nicholson and I. Shaw (eds.), *Ancient Egyptian Materials and Technologies*, 390–429. Cambridge University Press, Cambridge.

Serpico, M. and White, R., 2001, The Identification and Use of Varnishes on New Kingdom Funerary Equipment, in W. V. Davies (ed.), *Colour and Painting in Ancient Egypt*, 33–42. British Museum Press, London.

Sethe, K., 1907, *Urkunden der 18. Dynastie*, IV. J.C. Hinrichs'sche Buchhandlung, Leipzig.

Shaw, I., 2001, Egyptians, Hyksos and Military Technology: Causes, Effects or Catalysts?, in A. J. Shortland (ed.), *The Social Context of Technological Change: Egypt and the Near East, 1650–1150 BC*, 59–71. Oxbow Books, Oxford.

Shortland, A. J., 2001, Social Influences on the Development and Spread of Glass Technology, in A. J. Shortland (ed.), *The Social Context of Technological Change: Egypt and the Near East, 1650–1150 BC*, 211–222. Oxbow Books, Oxford.

Shortland, A. J., Nicholson, P. and Jackson, C., 2001, Glass and faience at Amarna: different methods of both supply for production, and subsequent distribution, in A. J. Shortland (ed.), *The Social Context of Technological Change: Egypt and the Near East, 1650–1150 BC*, 147–160. Oxbow Books, Oxford.

Smith, L., Bourriau, J., and Serpico, M., 2000, The provenance of Late Bronze Age transport amphorae found in Egypt, *Internet Archaeologist*, 9.

Spalinger, G. L., 1982, Stone Vessels, in *Egypt's Golden Age: The Art of Living in the New Kingdom 1558–1085 B.C.*, 126. Museum of Fine Arts, Boston.

Stern, B., Heron, C. P., Serpico, M. and Bourriau, J., 2000, A comparison of methods for establishing fatty acid concentration gradients across potsherds: a case study using Late Bronze Age Canaanite amphorae, *Archaeometry*, 42(2), 399–414.

Stern, B., Heron, C. P., Corr, L., Serpico, M. and Bourriau, J., 2003, Compositional Variations in Aged and Heated Pistacia Resin found in Late Bronze Age Canaanite Amphorae and Bowls from Amarna, Egypt, *Archaeometry*, 45(3), 457–469.

Taylor, J., 2001, Patterns of colouring on ancient Egyptian coffins from the New Kingdom to the Twenty-sixth Dynasty: an overview, in W. V. Davies (ed.), *Colour and Painting in Ancient Egypt*, 164–181. British Museum Press, London.

Tchapla, A., 2003, Analyse Physico-Chimique d'onguents et fards rituels trouves dans une tombe égyptienne, in M.-C. Grasse (ed.), *L'Égypte: Parfums d'histoire*, 152–161. Somogy éditions d'art, Paris.

Tylor, J. J. and Griffith, F. Ll., 1894, *The Tomb of Paheri*. Egypt Exploration Society, London.

Zeuner, F. E., 1963, *A History of Domesticated Animals*. Hutchinson and Co., London.

Zohary, D. and Hopf, M., 2000, *Domestication of Plants in the Old World*, 3rd Edn. Oxford University Press, Oxford.

Chapter 7

Minoan and Mycenaean Technology as Revealed Through Organic Residue Analysis

Holley Martlew

For John Evans, in memory

Abstract

This paper describes some of the more interesting results of organic residue analysis which were obtained in the Project, *Archaeology Meets Science*, the first part of which is completed, and the second currently underway, both initiated and directed by Dr. Yannis Tzedakis, Director General Emeritus of Antiquities, Greek Ministry of Culture, and the author. The discoveries include two workshops, one for herbal remedies at the Early Minoan III–Middle Minoan I (*c.* 2000 B.C.) site of Chrysokamino in East Crete and a workshop for aromatics dating to Middle Minoan IA (*c.* 2160–2000 B.C.), at the site of Chamalevri in West Crete. For evidence of use rather than production, medical preparations typical of skin balms were found at the Early Helladic Cemetery site of Kalamaki in the Greek Peloponnese (*c.* 3000–2700 B.C.). Of equal scientific importance, but of greater human interest, our chemists traced the history of Greek retsina; discovered the production of flavoured and medicated wines, as well as richly flavoured wines from other fruit; beer; the presence of honey and probably mead in Bronze Age contexts; and even a possible precursor of ouzo.

INTRODUCTION

The Project *Archeology Meets Science: Biomolecular and Site Investigations in Bronze Age Greece*, on which this paper is based, was initiated and directed by Dr. Yannis Tzedakis, Director General Emeritus of Antiquities, Hellenic Ministry of Culture, and the author, and was and is currently, being carried out by an international team of archaeologists and scientists. The object is the in-depth application of state of the art scientific analyses to a large body of ceramic artefacts and skeletal material from Bronze Age Greece. The results have been presented in an international exhibition entitled *Minoans and Mycenaeans: Flavours of Their Time*, which opened at the National Archaeological Museum in Athens, in 1999, and

Figure 7.1 Entrance to the exhibition Minoans and Mycenaeans Flavours of their Time, *National Archaeological Museum, Athens.*

has been mounted since, in an additional sixlocations: Rethymnon, Crete; Chicago, USA; Birmingham, UK; and Stockholm, Sweden; Naples, Italy; and Geneva, Switzerland (Tzedakis and Martlew 1999).[1] A volume is in preparation (Tzedakis, Martlew and Jones, forthcoming) which will include the primary scientific evidence.

The Project is still continuing and has been expanded. The first exhibition included material from ten sites. By the time the exhibition reached Stockholm in February 2003, it included 16 sites (Figure 7.2). The site descriptions in this paper are taken from those provided for the exhibition, by the excavators. The Project now extends from the Dodecanese and Cyprus to the Bay of Naples, and includes 28 sites. Our stated intention is to live up to the main title of our work, and to remain at the cutting edge, where *"Archaeology Meets Science."*

This paper was written in order to demonstrate the role organic residue analysis can play in deciphering the secrets of past technologies. The importance of such an investigation is that we have science on our side. We are not just guessing, and the theories we propose have some basis in fact. If organic residue analysis has resulted in the identification of a workshop, it is of paramount importance. In such cases the relevant tools might also be identified, even the way in which some of them might have been used. If organic residue analysis has resulted in the

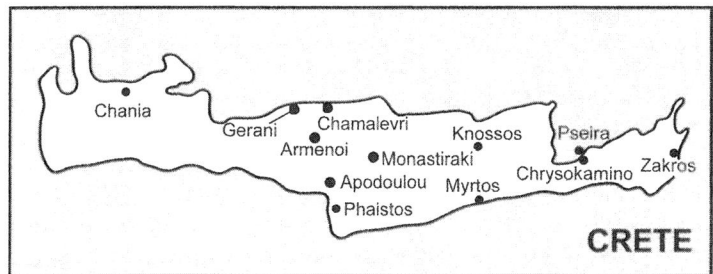

Figure 7.2 Map of Bronze Age Sites mentioned in the text.

identification of a product, it is of great significance. In both cases, the technologies can be assigned to their correct chronological periods, and so one may glimpse their beginnings, continuity, and development. Bronze Age Greece has proved to be a fortuitous subject for the purpose of demonstrating the role organic residue analysis can play in revealing the technological secrets of the past. Theirs was a wonderfully innovative world it seems, from the very beginning, to the very end. There was a great deal for us to discover and there still is.

EARLY TECHNOLOGY: THE EVIDENCE OF PRODUCTION

Chrysokamino (Early Minoan III/Middle Minoan IA, c. 2000 B.C.)

Chrysokamino is a small smelting workshop whose excavation was directed by Philip Betancourt, with the assistance of Jim Muhly. The site was in use from the end of the Final Neolithic until Middle Minoan IA (Betancourt *et.al.* 1999). The workshop consisted of a small hut with a soil floor, set on a slag pile that covered about 200 square meters. Finds were very fragmentary. They included pieces of furnace chimneys, pottery, stone tools, shells and animal bones, and pieces from more than 10 bellows, as well as fragments of slag and ore.

The smelting operation conducted at Chrysokamino was just one of the steps in the manufacture of metal tools and weapons. Because no ore was present near the site, it had to be mined elsewhere. The metallurgists used small bowl furnaces set in the ground. Clay chimneys helped the draft, and pot bellows pumped in a constant supply of air. The copper did not separate well from the other material in the furnace, but it formed small prills that could be removed by breaking the glassy slag. Some of this ore included small amounts of arsenic.

Most of the objects on the site come from its main period of activity, Early Minoan III/Middle Minoan IA (*c.* 2000 B.C.), and during this time a small structure was built on the slag pile.

This is what I said about Chrysokamino in a paper I gave three years ago:

> "The most prominent feature inside the hut is a hearth... Based on the potsherds found, Betancourt thought it likely that simple cooking was carried out, and that this little hut provided shelter for the workers from the violent winds that sweep the small exposed promontory. Now we are waiting to see what we can find out about life in a little Minoan industrial area in Early Minoan III." (Paper given at the conference *Minoans and Mycenaeans: Flavours of Their Time*, University of Bradford, December 2000).

We have subsequently identified some of the activities which took place there. According to Beeston and Beck, organic residue analysis has revealed that this small apsidal hut was not an area for cooking workman's lunches, but a workshop with a special purpose. The results indicate that the hut was used for the production of herbal remedies. These would have been used as medicines for treating occupational sickness. The arsenic content of the ore suggests the workers faced serious health hazards, specifically the results of arsenic poisoning, whose symptoms include skin changes, lesions (of palms and soles), respiratory inflammation, gastro-intestinal problems, and vascular wall injury.

At Chrysokamino, such symptoms would have been brought about by constant contact with arsenic-bearing copper ores. The smelting of the ores would have produced a highly toxic arsine gas and in metallurgical workshops such as Chrysokamino, exposure to low-dose arsenic would allow the substance to be absorbed and laid-down in bone, hair, teeth, and nails, with resultant chronic ill health. There would have been an attempt made to treat the symptoms, while the underlying cause would have remained unknown.

The workshop at Chrysokamino is a window through which one can see the creative way in which Minoans were apparently trying to cope with the health hazards of their time. This workshop is especially important because it predates the building of the first palaces of Crete.

All samples submitted to organic residue analysis are identified in this paper by Project number, followed by museum or excavation number (except in cases where museum numbers or excavation numbers were not assigned).

The analysis of the ceramic material from Chrysokamino was carried out by Professor Ruth Beeston. Organic residue results from Chrysokamino included:

– Project no. 524 (X 270) is a shallow bowl or cooking dish which contained a herbal preparation that included isophorone, camphor, and verbenone. Herbs include saffron, rosemary, sage, safflower, all of which have medicinal purposes. Also present were waxes that coat leaves, fruits, and petals of many plants, plant oils, and Aleppo pine which is native to Crete.

– Project no. 547 (X144), a jar which was said to contain a herbal medicinal preparation, the main contents of which were a variety of compounds found in rue, camphor and verbenone; and a compound which is present in fennel, cumin, and anise. Also found were alkanes, the most likely source of which was the waxy coating on leaves.

– Project no.527 (X1658), a closed vessel whose contents suggested a herbal preparation that included a variety of herbs, oil or fat, waxy leaves and resinated wine. It was pointed out by our chemists that wine and resinated wine were used for medicinal purposes in the ancient world.

– Another shallow bowl, Project no. 549 (X168), identified as having contained saffron. An open vessel, Project no. 525 (X525) was said to contain a herbal extract.

– A vessel of unidentified type, Project no. 526 (X685), with an organic residue result that included seven compounds of pine resin, aleppo pine, pine tar, and leaf wax. Saturated acids indicated a whole range of animal fats.

– Project no. 528 (X1850), a bucket jar; it revealed traces of licorice flavour found in fennel, cumin, and anise.

– Project no. 521 (X169), another bucket jar; it contained two unusual compounds that indicated species of thyme.

– Project no. 550 (X149) is a sherd which, it was thought, could have come from a bridge-spouted jar, and it gave an organic residue result of plant oils or wine, camphor, and a compound which might indicate oregano, saffron, and waxy coatings of leaves. Oil from wool was also considered a possibility.

Chamalevri, near Rethymnon, West Crete (Middle Minoan IA, (c. 2160–2000 B.C.)

In 1992 part of an open-air workshop was discovered at Bolanis, Chamalevri, near Rethymnon in West Crete. It was subsequently excavated by Maria Andreadaki-Vlasaki, Director of Antiquities for Central and West Crete (Vlasaki 1991–1993; 1994–1996). The site dated from the advanced MM IA phase, and there were strong indications that it was not a simple household workshop, but the site of an organised craft-industry of the period, but of exactly what type was difficult to conjecture. A long paved path afforded communication between, and also separated, what appeared to be individual activities. There were a large number of hearths and pyres of different types; clay vessels of a distinctive shape and specialised function; and large rubbish pits. The evidence made it possible to rule out the processing of metals and the production of pottery. There was undoubtedly widespread and intensive use of fire. The excavator concluded that the product in question was perishable. Such products could have been grain, olives, flax, and textiles. The discovery of a clay weight with the seal impressions of a floral motif directed thoughts provisionally to textiles and flax. I was told about the site, and decided it would be interesting to include Chamalevri in the Project.

There was another reason I wanted to include Chamalevri. One vessel, which was brought to me in fragments, looked like one of a group of enigmatic vessels that puzzled archaeologists, vessels whose purpose was unknown. This appeared to be one of them. They were called 'sheep's bells' by Sir Arthur Evans, because of their similarity in shape to a traditional sheep's bell.[2] I was excited at the prospect of solving a riddle that dated back to the beginning of Minoan archaeology, so I included a sherd from the so-called 'sheep's bell' in the group I sent to Professor Dr. Curt Beck for organic residue analysis.

Shortly after that, I had a communication from Beck to say that my special sherd, Project no. 105, had drawn a blank. The results were not simply organic signals which could not be assigned to a known commodity. There were no organic signals at all. Aside from my disappointment, this was not typical of samples in the Project which were averaging a 90% success rate in producing evidence of identifiable organic residues. It is the only time that I ever went back to a museum and asked to take another sample. What happened next was extraordinary. Not only did the second sample give chemical signals, there were 120 chemical components present. At the conclusion of the first part of the Project (the 'parent' project on which the exhibition was based), this vessel had revealed the highest number of organic signals of any of our samples. Not only that, but one of these components was identified as oil of iris. In today's perfume industry, oil of iris is the most expensive ingredient used in the production of perfume, selling for about 3000 pounds sterling, per kilogramme.

Beck produced a mass spectrum for acetoveratrone, the constituent of root of iris, for the vessel Project nos. 105/334 (RM 15312) (Figure 7.3). Beck's summary of the contents of this vessel was that it was a complex preparation for cosmetic use, which included oil of iris. Chemical components also included olive oil and pine resin. Other possible components (not confirmed by results as securely as was the oil of iris) included carnation and anise.

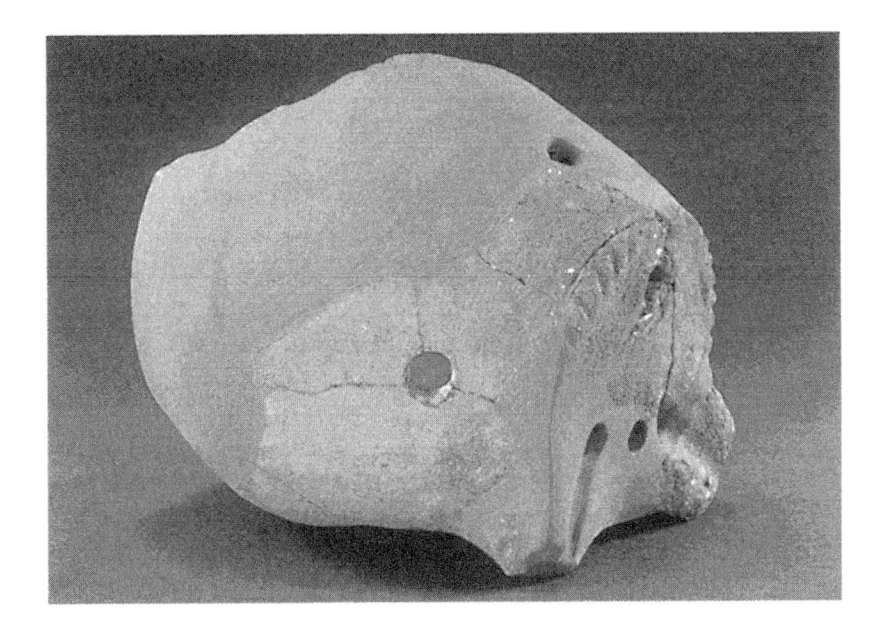

Figure 7.3 Industrial vessel (MMIA) Project Nos. 105 and 334, RM 15312. H. 21.5 cm (Catalogue no. 12).

Aside from satisfaction at the outcome, there was an important lesson to be learned. Organic signals are not homogeneous throughout a vessel. This is something we have constantly kept in mind since our experience with the industrial vessel from Chamalevri. It also turned out that when the vessel was restored, it was not a 'sheep's bell' after all, but a unique shape not seen before or since.

Another vessel found nearby was analysed and results showed it had also contained oil of iris and olive oil. This is the jar Project no. 106 (RM 15304) (Figure 7.4). A small phial, Project no. 339 (RM 15307) (Figure 7.5) produced residues identified as oil of iris, and a beeswax mixture with a strong cereal presence – the cereal presence could indicate its use as a thickening agent. The results of the analyses of samples from a total of five Chamalevri vessels (Project nos. 105/334, 106, 108, 335, 339 – museum numbers RM 15312, RM 15304, RM 15313, RM 15521, and 15307 respectively), suggested that this workshop was engaged in the processing of aromatic plants. The Linear B tablets from Knossos (Melena 1983), dating from the early 14th century B.C., in the Late Minoan period, afford incontrovertible evidence for aromatic oils. Now dating hundreds of years earlier, at the turn of the 2nd millennium, a tiny workshop in West Crete has given us the first scientific proof of aromatic production in the Minoan world.

One is led to the conclusion that some of the wealth that built the first palaces on Crete must have been produced by trade in rare and valuable products such

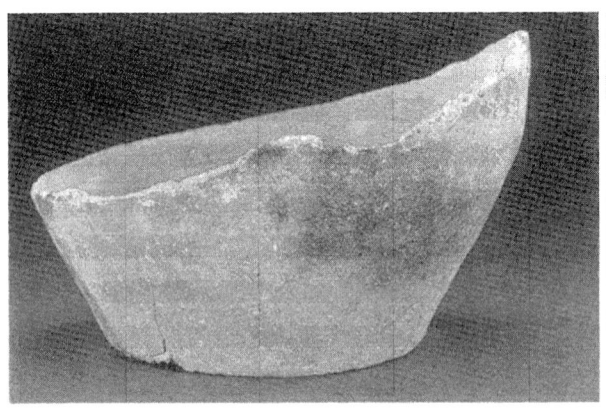

Figure 7.4 (left) Jar (MMIA) Project no. 106, RM 15304. H. (max.pres.) 14.5 cm (Catalogue no. 13).
Figure 7.5 (right) Miniature phial (MMIA) Project no. 339, RM 15307. H.3.9 cm (Catalogue no. 19).

as aromatics containing oil of iris, and that what we are seeing at Chamalevri is concrete evidence of the production of commodities which played a role in giving the Minoans their pre-eminence. "The most important things in life are food, clothing, and aromatic oils." This Assyrian quotation[3] gives an indication of the importance of aromatics in the ancient world. The temptation today is to think of aromatics as luxury goods whose primary use is cosmetic. In the Bronze Age aromatics and aromatic oils had a variety of important uses, including hygienic, medicinal, and ritual.

THE EVIDENCE OF USE

The Early Helladic Cemetery of Kalamaki (c. 3000–2700 B.C.)

The cemetery of Kalamaki is in the northwestern Peloponnese. The excavations are directed by Adamantia Vassilogamvrou. At a distance of 30 km southwest of the city of Patras, the cemetery occupies part of a low plateau at the foot of a mountain. During the Hellenistic and Classical periods the area was farming land, and so it continues to be in modern times.

Rescue excavations conducted in this area, after the illicit activities of antiquities dealers, revealed a cemetery of rock-cut chamber tombs, dating to the Late Helladic III period (c. 1600–1200 B.C.). It was then discovered that the Late Helladic tombs had been constructed at random over the site of a much earlier cemetery, one which dated to the Early Bronze Age (c. 3000–2700 B.C.). The early tombs had been partly preserved in the dromoi (corridors) and burial chambers of the later tombs. Occasionally an early tomb had been re-used as an ossuary in the Late Helladic period, and in one case, an early tomb had been re-used for a burial.

An Early Bronze Age rock-cut tomb consisted of a pit, or a dromos (corridor) and a small chamber, both dug into the soft rock. The dromoi gave access to chambers through small, and often very narrow, entrances, which were sealed after use, either by a slab or a dry stone wall. The tombs were used for multiple burials over an extended period of time. Three tombs had benches carved on the sides opposite the entrance. On the bench of Tomb 7, which was found undisturbed, there were skeletal remains that included some long bones, and a skull which had been placed next to a small flat stone like the 'pillows' of stone that have been found in tombs in the Cycladic islands in the Early Bronze Age (*c.* 3000–2000 B.C.). The indication is that a body had been laid on this bench.

The grave goods found in Tomb 7 are representative of the type of artefacts which these early people buried with their dead. The finds consisted mainly of pottery, but also present were two copper needles, a few amulets, flint arrowheads and flakes. All the vessels were handmade, and unevenly fired. Rock-cut tombs of this kind occur in many areas in the east and central Mediterranean, and they date from the late Neolithic to the end of the Early Bronze Age (*c.* 3000–2000 B.C.)

The most compelling reason for including the Early Helladic cemetery of Kalamaki in the project, was the possibility that it could provide us with the earliest evidence of what was used in burial ritual at the beginning of the 2nd millennium. The results did not disappoint us. The first scientific evidence from the Early Bronze Age in Greece, of compounds whose ingredients could be used in skincare, and therefore could have been used as embalming preparations, was contained in what could have been embalming bowls. The most important results are listed below. The primary evidence for the organic residue results, including those of year 2003, will be published in our forthcoming volume. In this paper I am including some of the primary evidence on one site, selected mass spectra for the organic residue results on five samples from the cemetery of Kalamaki, kindly supplied to me by the organic chemist who carried out the analysis of the ceramic material from Kalamaki, Dr. Vic Garner.

– Project no. 1107 (PM 4269) is a bowl found in the dromos (corridor) of Tomb 13. Might this have been an embalming bowl? It was described as having contained a medical preparation typical of a skin balm treatment (Figure 7.6). According to Garner, the GC/MS analysis of sample 1107 identified an almost symmetrical profile of long-chain hydrocarbons/waxes (C24-C30) with significant contributions from higher terpenes and sterols. These features can be seen in the Mass Chromatograms reproduced in Figure 7.7. This approach of viewing mass chromatograms of diagnostic ions instead of using the Total Ion Chromatograms affords a convenient means of identifying profiles of chemicals that are recognisable either in terms of origin or mode of use. In this case, such a mixture of hydrocarbons, waxes, terpenes and sterols constitutes a perfumed oil and one likely to have topical medical properties. It is similar to several modern day cosmetic formulations used for skin treatments.

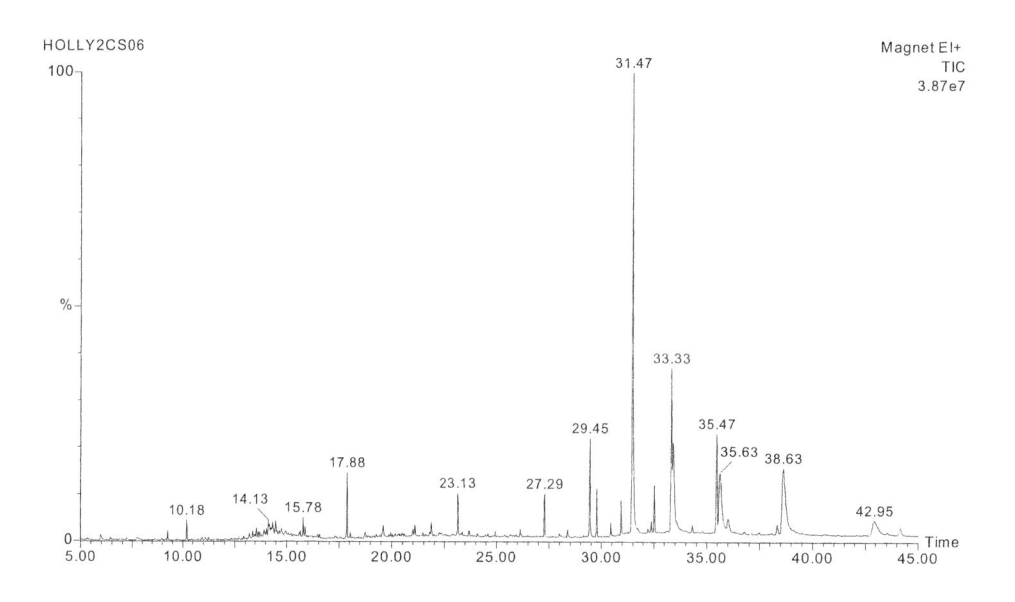

Figure 7.6 Project no. 1107, PM 4269. Total Ion Chromatogram.

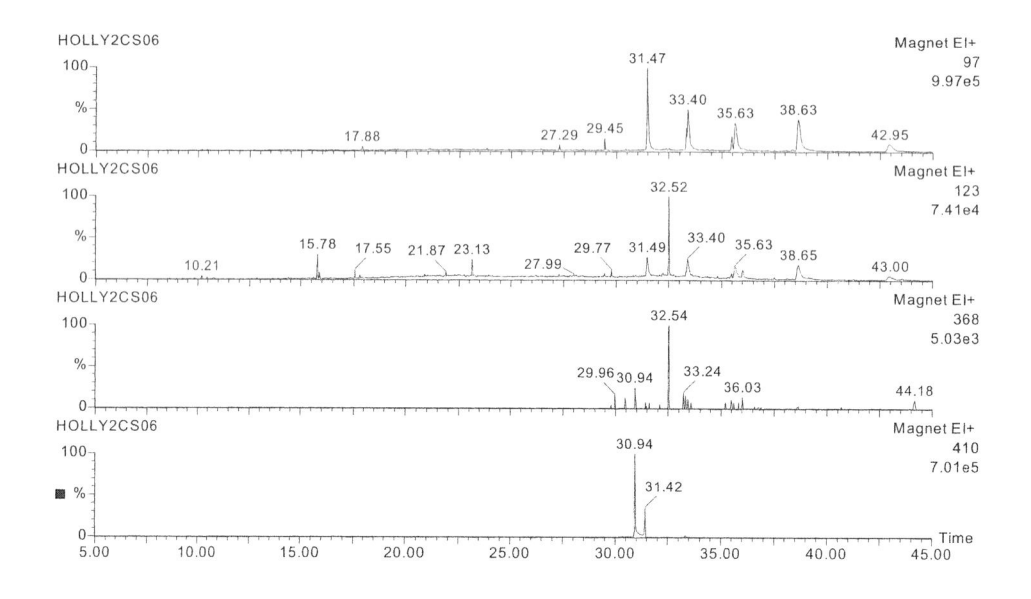

Figure 7.7 Project no. 1107, PM 4269. Mass Chromatograms.

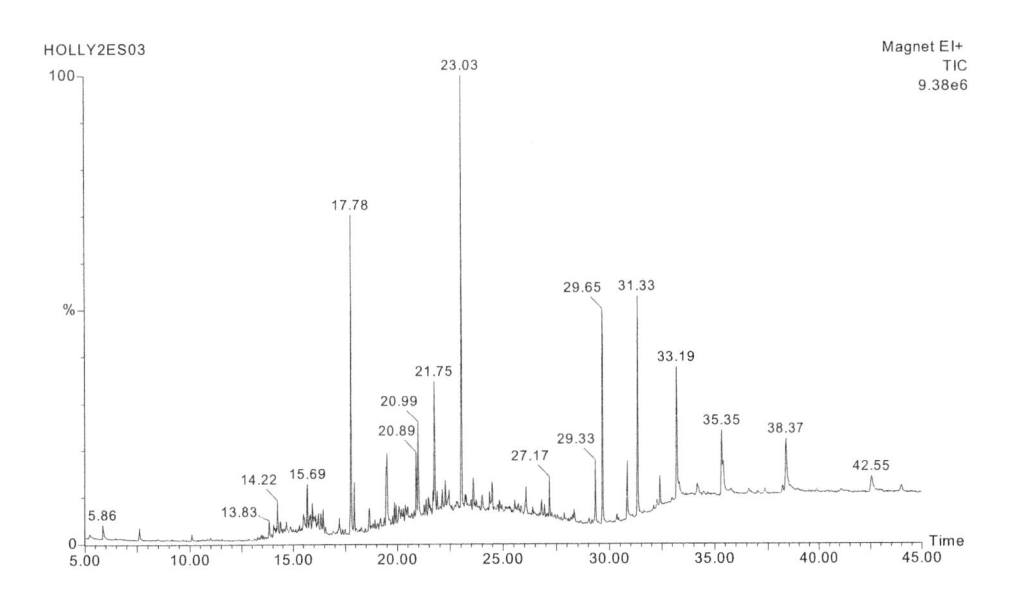

Figure 7.8 Project no. 1117, PM 14277. Total Ion Chromatogram.

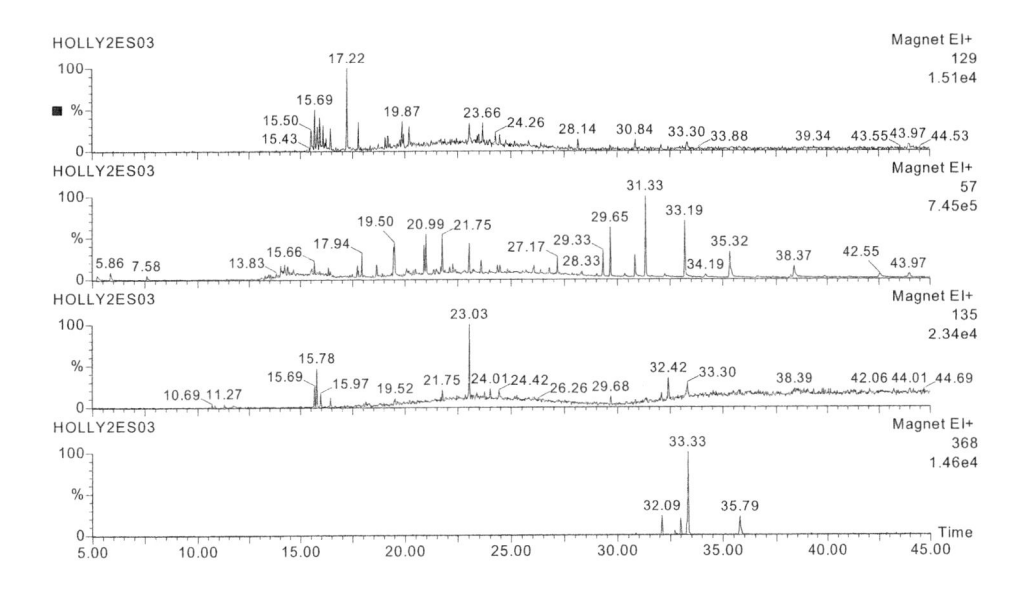

Figure 7.9 Project no. 1117, PM 14277. Mass Chromatograms.

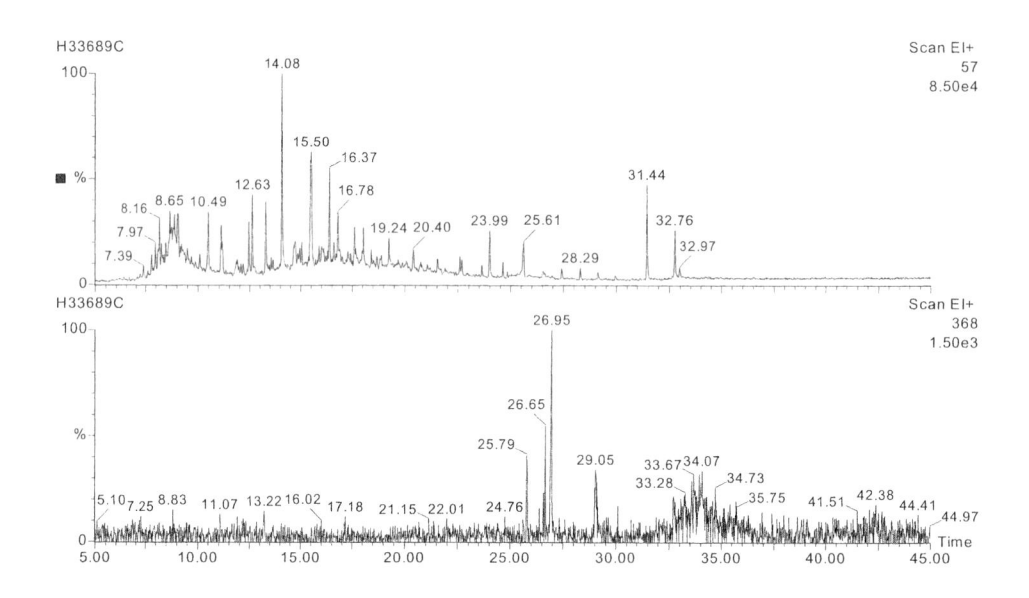

Figure 7.10 Project no. 774, PM 14278. Mass Chromatograms.

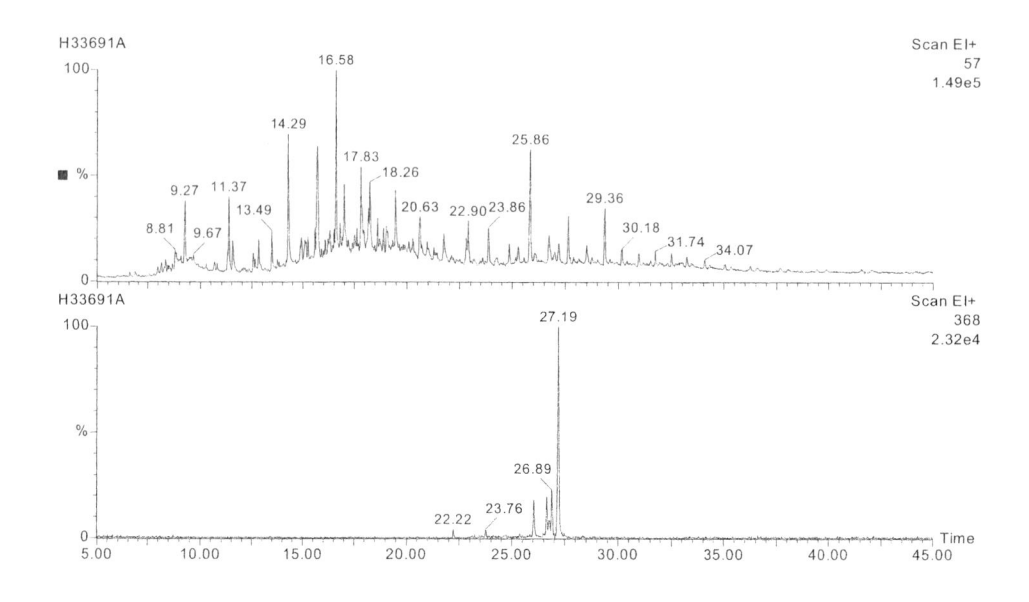

Figure 7.11 Project no. 790, PM 14280. Mass Chromatogram.

– Project no. 1117 (PM 14277), from the dromos (one corridor for 2 tombs) of Tombs 1.3 and 1.3a, is another example of a vessel that could have been an embalming bowl. This vessel contained a complex ointment or balm, which could have been used in an embalming process. The mass chromatograms (Figures 7.8–7.9) show complex profiles of hydrocarbons, fatty acids and terpenes including cholesterol derivatives. Their presence is taken as indicative of an animal origin for sample 1117.

Three vessels from Kalamaki are cited as having contained animal fat preparations which were the result of rendering or heat treatment. The indication was that they could have been unguents prepared for funerary ritual. These are Project nos. 774, 790, and 791.

– Project no. 774 (PM 14278) comes from the wall of a vessel of unknown type found in the dromos (corridor) of Tomb 48 (Figure 7.10).

– Project no. 790 (PM 14280) is a bowl found in the dromos of Tomb 7 (Figure 7.11). Garner explained that the branched chain hydrocarbons pristine (C_{19}) and phytane (C_{20}) can be identified in the GC/MS data. The ratio of these two compounds is used to distinguish plant and animal origins: usually in plant derived samples $C_{19}<C_{20}$ whereas animal origins generally show $C_{19}>C_{20}$.

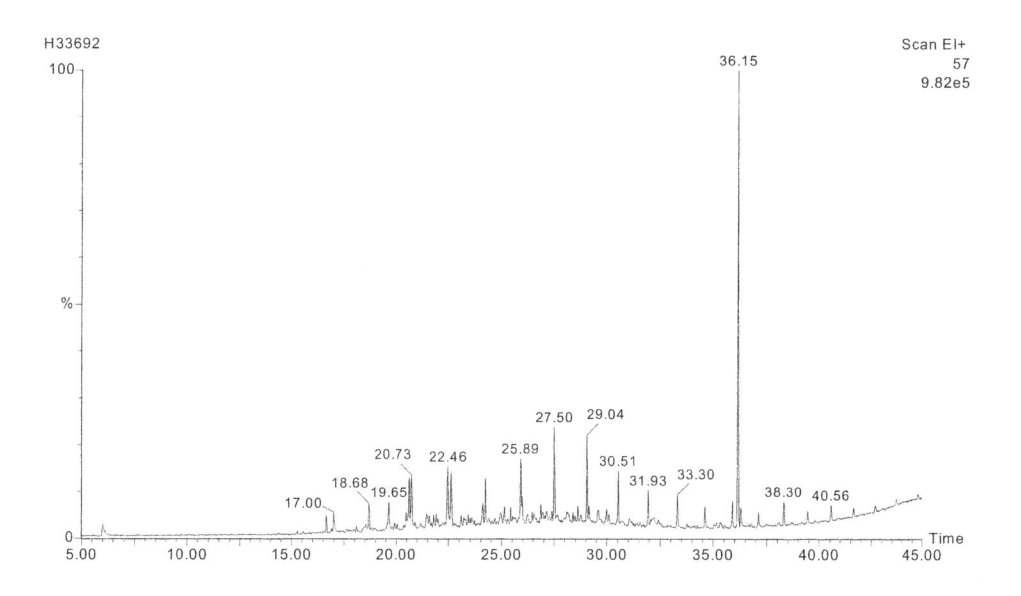

Figure 7.12 Project no. 791, PM 1255. Mass Chromatogram.

– Project no. 791 (PM 1255) is the base of a vessel whose type is unknown, which was also found in the dromos of Tomb 7. The Mass Chromatogram (Figure 7.12) (m/z = 57) shows high carbon numbers and a wide range in the hydrocarbon profile indicating degraded animal fats. The strong signal at Rt = 36.15 is assigned to squalane arising from incautious handling of the specimen (fingerprint fats).

The Late Helladic Cemetery of Sykia (Late Helladic IIIA–IIIB, c. 1390–1185 B.C.)

The Mycenaean cemetery of Sykia is situated near the village of Sykia about 70 km southeast of Sparta, and about 10 km from Monemvasia in the very southeast prong of the Peloponnese. The Late Helladic sherds which gave the first indication of a Mycenaean site were identified by the British School at Athens. The cemetery itself was discovered in 1972. Four tombs have been excavated under the direction of Yanna Efstathiou. It is clear that there are many more tombs in the area. All of the work carried out to date has been generously funded by the Mayor of the village of Sykia.

The four tombs, all oriented from northwest to southeast, were cut into the local serpentine. Each of them consists of a dromos or corridor and a chamber. All four contained multiple burials of what are thought to be family groups. As appears to be the pattern in such tombs, the last burial is preserved *in situ* on the floor of the chamber. In most cases, bones mixed with offerings were pushed aside in order to make space for new burials.

There are three important aspects of the results of the organic residue analysis of the contents of vessels from the cemetery of Sykia: the first is the method of non-destructive sampling which we are pioneering in order to sample intact vessels; the second is how the results inform us about burial practices in the Late Mycenaean world, *c.* 1390–1185 B.C.; and third, the unusual compounds which have been identified.

Five of the vessels cited below are complete vessels, without restoration. These were analysed using the non-destructive method. Included are the piriform jar and the two cups referred to below.

– Project no. 350 (SM 13034) is a piriform jar found in the chamber of Tomb 2. The contents of this vessel suggested incense whose use is discussed below.

– Project no. 357 (SM 1301) is a small spouted cup found in the chamber of Tomb 2. Animal fats in the cup indicated the likelihood that the contents had been an embalming oil. The long spout and side handle on this decorated vessel show a need for the controlled and delicate pouring of a special (precious) substance.

– Additionally, the contents of another spouted cup (Project no. 360; SM13008) and a kylix/goblet (Project no. 388; Sparta Museum 13017) indicated herbal extracts and the presence of water.

– Project no 371 (SM 13025) is a sherd from a coarse vessel that had contained residues of honey or a preparation sweetened with honey.

A fingerprint and a modern antiseptic is the organic residue result for Project no. 380 (SM 13031) which is a sherd of a painted vessel found in the dromos/corridor of Tomb 4. This result clearly illustrates the type of detail that organic residue analysis is now capable of giving and the skill and care required on the part of the analyst.

WHAT DID THEY DRINK?

The subject of the Project is food and drink, but in a short paper, why describe stews of pork, lentils, and olive oil, when one can describe cocktails of resinated wine, barley beer, and honey mead? The two drinks most associated with modern day Greece are retsina and ouzo, so I shall discuss these first, but in reverse order to retain the correct chronological sequence.

Ouzo?
The Early Helladic Cemetery of Kalamaki
– Project no. 1111 (PM 14271) is from the wall of an open clay vessel from the dromos/corridor of Tomb 42. The organic residue analysis identified 'brandy lactones.' A flavoured liquor and camphor were also present. Garner suggested that this mixture could have been a precursor to ouzo. The project had previously traced the history of retsina. Now it appears that Greeks might have been drinking a type of ouzo *c.* 3000–2700 B.C.

Retsina

Retsina is wine to which pine resin has been added. Greece is the only modern country to produce such resinated wine, but no one had any idea when or where the practice of adding pine resin to wine originated. (I am making a distinction between the use of pine resin in wine, as opposed to other resins such as terebinth.). The combined results of the parent Project and the new one have traced the history of retsina in Bronze Age Greece, and found evidence dating from Early Minoan III (*c.* 2000 B.C.) to Late Helladic IIIC middle (*c.* 1090 B.C.). The first identification came from the Late Helladic period but slowly we worked our way backwards, to Late Minoan III, to Late Minoan I, to Middle Minoan II, until we confirmed the use of pine resins in a vessel dated to Early Minoan III. It was by chance that the results came in reverse order, but it made the whole investigation much more exciting.

We were amazed with the first discovery: a cooking pot jar found in the 'Room with the Fresco,' the Cult Centre at the palace of Mycenae, with a date of LHIII B2 (*c.* 1250 B.C.), had contained true retsina – wine with pine resin. The vessel is Project no. 68, MM 24350 (Figure 7.13). At a height of 34 cm and with a maximum diameter of 38 cm, it is large for a cooking pot found on the Mycenaean Mainland. The size of this vessel, along with the other evidence for wine found in vessels from the Cult Centre which were analysed, indicates the quantity of wine that must have been used/consumed as part of Mycenaean ritual. I am particularly

Figure 7.13 Cooking jar (LHIIIB) Project no. 68, MM 24350. H. 34 cm (Catalogue no. 178).

happy with the fact that this was our first find of true retsina, since I rescued the fragments of this pot from several bags of sherds and subsequently made the decision to send one for analysis.

The Middle Minoan II (*c.* 1700 B.C.) evidence for wine with pine resin came from a tripod cooking pot, Project no. 30, from the palatial centre at Monastiraki, in the Amari Valley in West Crete.

The sherd of a jar from Pseira, Project no. 543 (PS 219), dated to Early Minoan III (*c.* 2200–2160 B.C.), gave an organic residue result of seven pine resin acids. A short site description of the island of Pseira is presented later in the paper.

Terebinth Resin

A sherd from a small conical bowl, Project no. 148, from the Middle Minoan site of Apodoulou (*c.* 1700 B.C.) (Godart and Tzedakis 1992) gave an organic residue result of terebinth resin, *Pistacia atlantica* (turpentine tree), identical to the sample KW 165 from the Uluburun shipwreck. Beck has provided the following comment on *P. atlantica* and *P. terebinthus* resins:

"The *Pistacia atlantica* vs. *Pistacia terebinthus* is a very knotty question. The compound we found (in Apodoulou Project no. 148) is identical with one from the resin of the Uluburun (=Kas) shipwreck that was identified by Mills and White as *P.atlantica*. But *P. atlantica* is the only Pistacia species on whose resin extensive work has been done. I am now collecting references on all Pistacia species from Chemical Abstracts. There are hundreds of them. One says that *P. atlantica* and *P. terebinthus* hybridize naturally. Margaret Serpico's excellent summary in the new edition of '*Ancient Egyptian Materials and Technology*' (2000) does not list *P. atlantica* in Crete, only *P. terebinthus*, and that may be where you and Peter Warren get your information. But the Uluburun wreck clearly shows that *P. atlantica* resin was traded in huge amounts throughout the eastern Mediterranean, including to Egypt, where neither species occurs naturally. At the present time, there is no way to distinguish *P. atlantica* and *P. terebinthus* by chemical analysis of their resins." (Curt Beck, pers. comm., August 2002).

Copals

The sherd of a side-spouted jug, Project no. 514 (PS 4573), dating to Middle Minoan I-II (*c.* 2160–1700 B.C.), from Room 2, Building BC on the island of Pseira, was analysed by Beck. The summary is: pine resin; animal fat; plant waxes and/or beeswax. Two resin acids appear to be copals, which are hard, translucent, odoriferous resins obtained from various tropical trees which do not grow in Crete. Africa would be a source. Could there have been a trade in copal resins at this early date? Why not, if *P. atlantica* was already being traded?

Mastic?

A decorated piriform jar, Project no. 346 (SPM 122986), from the Late Helladic cemetery of Sykia in the Peloponnese was analysed for organic residues. The full result identified a tree resin, like mastic. The evidence, as reported by Garner, indicated that the vessel contained dried flakes of resin which would have been collected from the tree, cracked and kept either to flavour wine, or to be burned, perhaps by being thrown on an open fire. What gives this find further interest is that the shape of the jar is such that it would have been top heavy if filled with a liquid, but filled with something light, like dried flakes of resin, it would have kept in balance. Our research could be throwing light on uses of unusual (and possibly unwieldy, however beautiful) Mycenaean vessel shapes.

Aromatic Wines

Mycenaeans in the West: Vivara, Bay of Naples (Late Helladic I, c. 1600–1519 B.C.)

The island of Vivara is in the Bay of Naples, between the islands of Procida and Ischia, and is the remaining western portion of a volcanic crater. The island has a surface area of just 0.32 sq. km but it is in a strategic location and is naturally fortified by high cliffs which enable the island to control sea routes. In the 17th century B.C. it was a densely inhabited area on the metal trade routes between the eastern and western Mediterranean. Copper came from the Thyrrenian mining areas, mainly in Tuscany and Sardinia, and was collected at Vivara by people from the Aegean. The emergent Mycenaean Greeks had a great need for metals. They used them to display wealth and status as evidenced by the weaponry in the shaft graves at Mycenae. Archaeological finds on Vivara illustrate constant contact with the Peloponnesian world (Marazzi 1998; Marazzi and Tusa 2001).

In the settlement, there was a great deal of evidence of smelting, as well as recycling, of metals. Fragments of melting crucibles, slag, prills, melting residues, a mould, and copper and bronze droplets were found, all confirming the importance of the island. A large number of vessels was found, including finely decorated vessels of Mycenaean Greek origin. Helladic domestic ware was also found, most of which was hand-made, and of a type that was not normally exported from Greece.

The vessels of Vivara give us a rare insight into the diet of the Mycenaean Greeks who travelled. Wine flavoured with herbs drunk from an elegant Vaphio cup. The organic residue result summary reads "herb flavoured unresinated wine" (the Vaphio cup is Project no. 451; ISOB V95e/61).

Because of the special interest a Mycenaean presence abroad invokes, mention must be made of other organic residue results on finds from this site: a vegetable oil with a wood herb bark (as opposed to a leafy herb like parsley) was found in a Canaanite jar (Project no. 447, ISOB VE82e/360). A small Mycenaean askos (jug) delicately painted in a crocus pattern, had held an olive oil flavoured with a herbal extract (Project no. 463; ISOB V257). A domestic vessel, a barbotine jug, which has been restored, contained residues that indicate a range of herbs which could have been used to flavour milk or cream (Project no. 434; ISOB V2001C/12). This is a unique result for the Project, and invites comment on the ingenuity of the cooks who were living at Vivara. Flavoured butters or creams indicate both culinary skill and refined taste. The ladle or cup, found on the site, had contained herbs (Project no. 429; ISOB VE01C/7).

The picture presented at Vivara is of a cultured people who insisted on a high standard of living when they were residing in a foreign land. The opulence of the burial goods in Grave Circle A at Mycenae, dating to the 16th century B.C., present a similar picture of a people with highly developed tastes.

FLAVOURED AND POSSIBLY MEDICATED WINES

Pseira, a tiny islet near the northeast coast of Crete

Geologically, Pseira is part of Crete. It was inhabited during the Bronze Age, and it supported a substantial town. It was abandoned at the end of the Bronze Age, and it remained uninhabited until the Byzantine period. The town on Pseira Island was excavated in two campaigns; in 1906–1907, by Richard Seager and a second period of excavations between 1985 and 1992 was conducted by a Greek-American team, under Costis Davaras and Philip Betancourt. The second campaign revealed additional details and made Pseira the most completely studied Minoan town in Bronze Age Crete. Organic residue analysis is adding a further dimension to our understanding of the site.

Seven vessels from Pseira contained resinated wine, and four of these contained flavoured resinated wine. Three of these vessels contained a range of aromatic botanicals and two of these vessels appear to have contained rue. The evidence, according to Beck who carried out the analysis of the vessels from Pseira, points to the possibility that what we are seeing is not just a flavoured wine, but a medicated wine. What makes this evidence even more interesting are the dates of the vessels that could have contained a medicated wine.

These vessels from Pseira date from Middle Minoan (c. 2160–1600 B.C.) to Late Minoan IB (c. 1480–1425 B.C.), and as such, indicate a continuity of production and provide evidence for resinated, flavoured, and possibly medicated wine:

– Project no. 516 (PS 4579) is an open vessel found in Building AF, Room 7, and dated to Middle Minoan (c. 2160–1600 B.C.). The organic residue result was wine resinated with the resin of the Aleppo pine and flavoured with rue and possibly other aromatic botanicals. What is indicated is either a flavoured, or a medicated wine.

– Project no. 532 (PS 1075) is a basin found in Building BS, and dated to Late Minoan I (c. 1600–1425 B.C.). The organic residue result was a wine that is flavoured with pine resin, and a range of other botanicals, that could include rue. Perhaps the aromatics were only for flavour, but they could have been for medicinal use.

– Project no. 515 (PS 4582) is an open vessel, found in Building AF, Room 7. It dated to Late Minoan I (c. 1600–1425 B.C.). The organic residue result was three pine resin acids. The most remarkable aspect of the analysis is the presence of either natural bitumen (coal tar), or the degradation products of fats and oils. It remains an open question.

– Project no. 511 (PS 4594) is an open vessel found in Building AF, Room 6. It dated to the Middle Minoan period (c. 2160–1600 B.C.). The organic residue result was a flavoured, resinated wine.

Salamis (Late Helladic B–C early, c. 1200 B.C.)

The sherds that were analysed to produce the results from Salamis have a special value in that they have helped provide the proof that there had indeed been a prehistoric settlement on the island of Salamis. Among the larger islands of the Saronic Gulf, Salamis lies nearest to Attica. Its fame derives mainly from the great sea battle that took place in the Straits in 480 B.C., when a united Greek fleet destroyed a much bigger Persian armada. The history of Salamis has now been enriched with the knowledge that the island also played a significant role in the Mycenaean World of the 2nd millennium B.C. The excavations are directed by Yannos Lolos.

– Project no. 748 (PAM Exh. No. 5) is a kylix or pedestalled goblet from Mycenaean Salamis which gave the organic residue result of a richly flavoured wine, like blackberry wine or *creme de cassis*. This is a unique result to date.

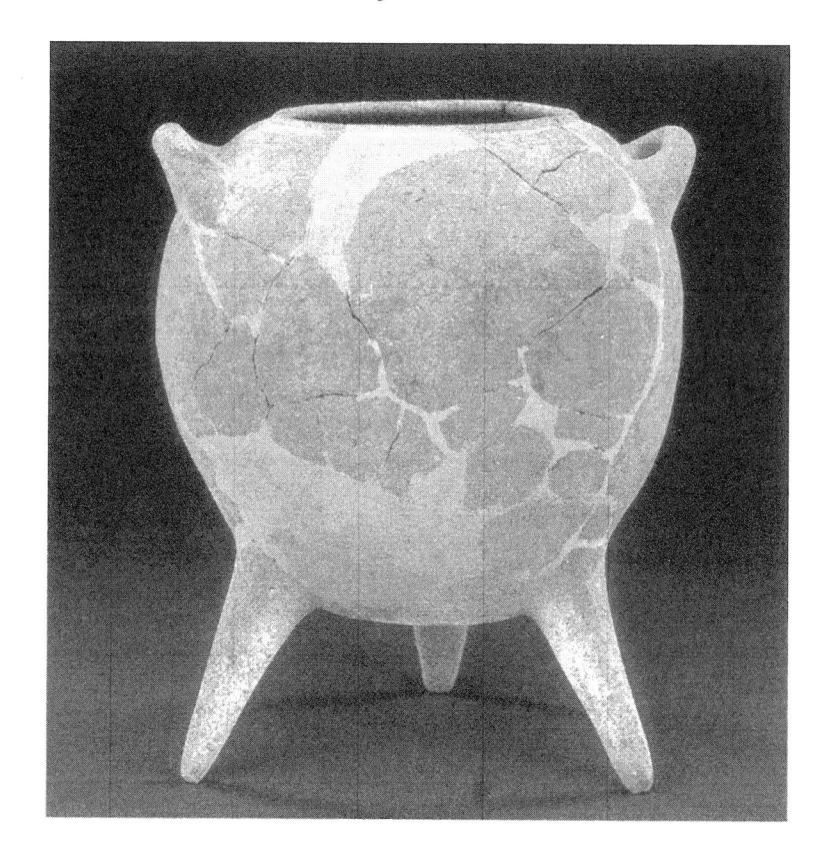

Figure 7.14 Tripod cooking pot (MMIIB) Project no. 21, RM 4747. H. 32.7 cm (Catalogue no. 54) Substitute.

Beer

Two organic residue results of Beck give indications of brewed beer. They are both tripod cooking pots, and both came from the Middle Minoan site of Apodoulou in the Amari Valley, southwest Crete (Godart and Tzedakis 1992). Both finds date to Middle Minoan II (*c.* 1700 B.C.).

– Project no. 21 contained phosphoric acid. This acid has been found in Predynastic (earlier than 3000 B.C.) beer brewing vats in Hierakonapolis, Upper Egypt (Tzedakis and Martlew 1999, 192). As this acid is a component of all living cells, a definite conclusion cannot be drawn. However results also produced 2-octanol. This is commonly thought to indicate beer (Figure 7.14 illustrates the type).

– Project no. 22 contained phosphoric acid, as did the one above. It also contained dimethyl oxalate, which is commonly thought to indicate beer.

Figure 7.15 Mug (LHIIIA2)) Project no. 195, MM 8011. H. 14.8 cm (Catalogue no. 157) Substitute.

Honey, possibly Mead
 – Project no. 195 from the palace of Mycenae, is dated Late Helladic III A2 (*c.* 1370–1340 B.C.) (Figure 7.15 illustrates the type).

The organic residue result on this vessel said it had contained wine and honey. The honey content indicated, according to Beck, the possibility of mead. This type of vessel has always been referred to as a mug, and sometimes a beer mug, because of its shape. Some of them might have been used for beer but not apparently this one.

Highs and Lows
The following tripod cooking pot yielded another interesting result. This type of vessel, with a cruciform leg section, has been found at various Mycenaean sites, including Mycenae and Thebes.

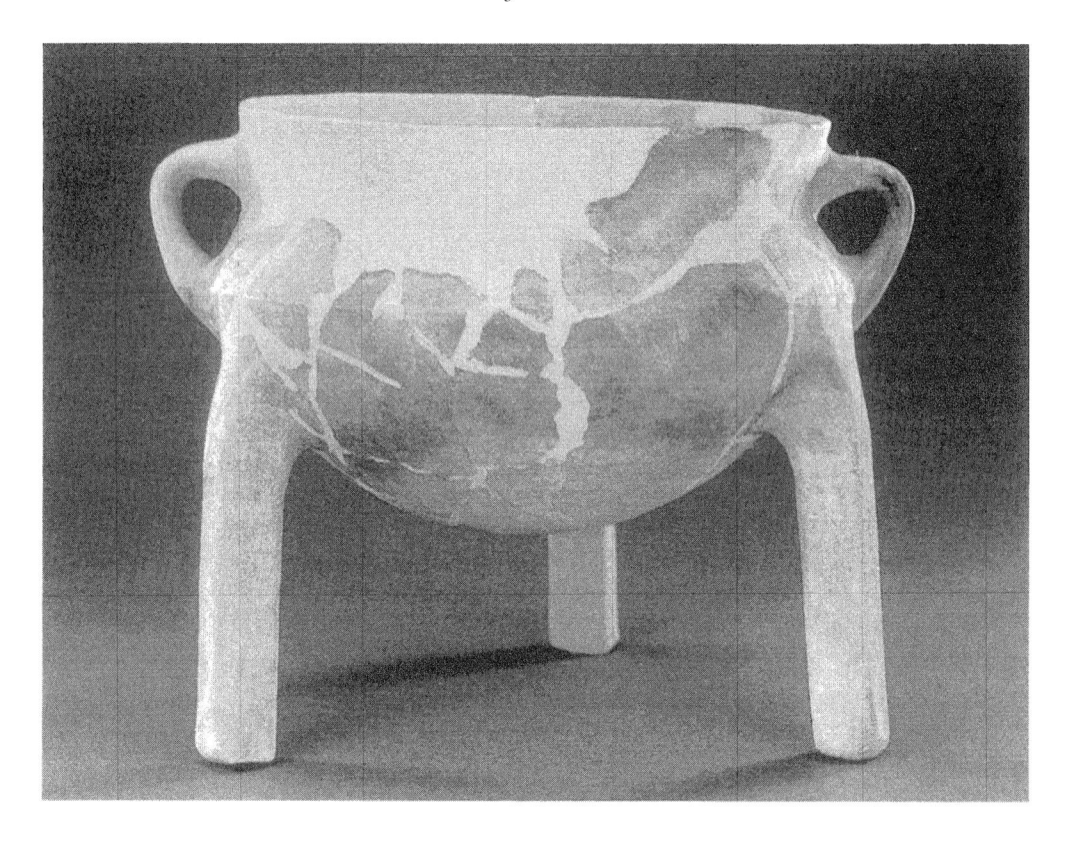

Figure 7.16 Tripod cooking pot (LHIIIC early) Project no. 72, MM 24323. H. 22.8 cm (Catalogue no. 148).

– Project no. 72 (MM 24323) (Figure 7.16) from the 'Citadel House,' Room xxxiv at Mycenae. It dates to Late Helladic IIIC early (c. 1185–1130 B.C.). The organic residue result according to Beck was resinated wine with the herb rue (no other herbs present). Rue is a mild narcotic and stimulant. So what did the Mycenaeans have in mind when they had drunk this potion? Or rather what state of mind were the Mycenaeans in after they drank this potion? Project no. 547 (X144), from Chrysokamino (see above) was cited as having contained rue.

The Minoan Cocktail: Resinated Wine, Beer, and Honey Mead

The organic residue work undertaken by Beck, Garner and Beeston has played a predominant role in this paper, but I can hardly speak of drink without a reference to the 'cocktail' proposed by Patrick McGovern (Tzedakis and Martlew 1999, 167, 176) of resinated wine, beer and honey mead as being the contents of several vessels, including the two types above. The references are:

Figure 7.17 Conical cups (MMIA) Project nos. 36,47,61. RM 16002, 16004, 16006, 16007, 16009, 16015, 16018, 16026. H. 5-6.2 cm (Catalogue no. 25) Substitutes.

– Project nos. 36, 47, and 61 from Chania Splanzia, are conical cups, dating to Late Minoan IA (*c.* 1600-1480 B.C.) (Fig. 17 illustrates Minoan conical cups).

– Project no. 111 (RM 17284) is a sherd from a kylix/goblet from the Late Minoan cemetery of Armenoi, Tomb 178, Late Minoan IIIA2 (*c.* 1370–1340 B.C.) (Figure 7.18 illustrates the type).

The primary reason for thinking this was a cocktail, and that the drinks were not imbibed sequentially is because, in certain cases such as these, they were found in ritual contexts: the conical cups from a foundation deposit, and the kylix from a ceremonial pit at a cemetery. The mixture seemed strange when first proposed, but drink is subject to fashion, and we have since discovered that, in Scandinavia at a very early date, a mixture of fruit wine, mead and beer was being drunk. A visitor from Nebraska who came to the opening of the exhibition at the National Archaeological Museum in Athens made the comment that mixing wine and beer was still common in certain areas of the American mid-west.

STORAGE CONTAINERS

A priori, it would seem unlikely that residue analysis would contribute evidence relating to methods of storage. However, the following analyses yielded some interesting and suprising results. At the Early Helladic cemetery of Kalamaki in the Peloponnese, as previously described, the wall of a vessel, Project no. 799 (PM 14287), found in the dromos/corridor of Tomb 10, was said by Garner to have contained residues from liquors that had been stored in wooden/oak casks.

Dating to Middle Minoan (c. 1700 B.C.), at the palatial centre of Monastiraki, Amari Valley, West Crete, the organic residue result on a sherd from a tripod cooking pot, Project no. 30, was that it had contained "resinated wine that was stored in an oak barrel, that may well have been a toasted oak barrel," according to Beck and McGovern. The oak lactones would increase with charring and this would increase the likelihood of detection. Toasted oak would give the wine a distinctive taste, such as is enjoyed by scotch whiskey drinkers today. The alternative intrepretation is that toasted oak chips had been added to the wine. Beck and McGovern commented in their summary of the results that ancient shipbuilding technology probably also involved the use of toasted oak timbers.

From the island of Pseira, Late Minoan I (c. 1600–1425 B.C.), one of the vessels, a jug with painted decoration, Project no. 512 (PS 4588) from Room 6, Building AF North, was analysed by Beck, who reported that it had contained wine

Figure 7.18 Kylix (LMIIIA2) Project no. 111, RM 17284. H. 11.8 cm (Catalogue no. 167) Substitute.

Figure 7.19 Rhyton (LHIIIA2) Project no. 330, MM 24362. H. 27.3 cm (Catalogue no. 164) Substitute.

preserved either by the deliberate addition of pine tar, or was wine that had previously been kept in a vessel lined in pine tar. Another of the vessels, Project no. 515 (PS 4582) discussed above, could have contained pine tar. The reasons for the presence of pine tar is still being investigated.

Finally, I would like to mention two Minoan vessels, whose uses/contents have been the subject of much discussion for many years. Both results are from John Evans, which is a fitting and proper way to come to the end of a paper dedicated to the memory of a valued friend and revered colleague. There is an example, Project no. 330 (MM 24362) (Figure 7.19 illustrates the type) of a Minoan ritual vessel called a rhyton, thought to have been used for libations. Libations of what? Evans' analysis of the residues revealed barley beer and wine. The original rhyton came from Midea, a fortified site near Mycenae. It dated to Late Helladic IIIA2-IIIB (*c.* 1370–1190 B.C.).

There is another vessel from Midea, Project no. 331 (MM 28092 (Figure 7.20). The illustration shows the actual vessel which was tested, and it dates to Late

Figure 7.20 Feeding bottle (LHIIIB) Project no. 331, MM 28092. H. 7.2 cm (Catalogue no. 158).

Helladic IIIB (c. 1340–1185 B.C.). Vessels of this shape have historically been called feeding bottles. Residue analysis revealed traces of beeswax – honey ...fermented product. Possibly mead or beer or a mixture of both . Can this be what babies imbibed in Bronze Age Greece?

SUMMARY

Our results have significant and surprising implications for our understanding of the lifestyle of the people living in the Minoan and Mycenean world in the period immediately before the building of the first palaces. The use of oil of iris was completely unexpected; the hypothesis that the hut at Chrysokamino, an Early Minoan smelting site, was a workshop for herbal remedies could never have been advanced without the evidence of residue analysis.

The most astonishing thing revealed by the totality of the results presented: from the embalming ointments found at the Early Helladic cemetery of Kalamaki; to the aromatic workshop at Pre-palatial Chamalevri in Crete; to the range of alcoholic potions enjoyed(!) by the inhabitants of Greece, from the Early through to the Late Bronze Age, is the sophisticated tastes and practices of these people, from the beginning to the end of their civilisation. What I find most exciting, however, is the very early evidence and so I have gone beyond the scope of most of the papers in this volume in order to bring it to the reader.

It is equally clear from the totality of the results which we have had so far, that there was an overwhelming use of herbs for food, for drink, for medicine and for ritual. It is equally clear that, like most of us today, they enjoyed a drink.

Their creativity in pursuit of a standard of living and a standard of dying, is endlessly fascinating. Their achievements at such an early date are impressive and a lesson to us. Even five years ago, it would have seemed preposterous that we could be discussing the possibility of toasted oak barrels or toasted oak chips in wine, resinated with pine no less, in 1700 B.C. and in Crete? But, even though we appear to have found out so much, this is only the beginning. Many results and interpretations will change, but I remain excited to this day, by every single result, by every single thing we find out about the inhabitants of Greece in the Bronze Age.

ACKNOWLEDGEMENTS

This paper has been written in memory of John Evans. I wish to acknowledge the support and collaboration and offer my very warm thanks to Dr. Yannis Tzedakis, Director General of Antiquities Emeritus, Hellenic Ministry of Culture, without whom none of this research or this Project could have taken place. I also give grateful thanks to Dr. Vic Garner for his valuable contribution to this paper. I wish to acknowledge and thank the other Project scientists whose work has been cited in this paper: Prof. Dr. Curt W. Beck, Prof. Ruth Beeston, and Dr. Patrick McGovern; and the Project excavators whose sites are included in this paper: Dr. Maria Andreadaki-Vlasaki, Prof. Philip Betancourt, Prof. Costis Davaras, Yanna Efstathiou, Dr. Elizabeth French, Dr. Claudio Giardino, Prof. Spyros E. Iakovidis, Dr. Yannos Lolos, Prof. Massimiliano Marazzi, Dr. Christina Merkouri, Dr. Carla Pepe, Dr. Adamantia Vassilogamvrou, and Prof. Gisela Walberg, and one of the Project members to whom I am indebted in many different ways: Dr. Robert Arnott. Special thanks is also due to the Istituto Universitario Suor Orsola Benincasa of Naples and the Soprintendenza archeologica per le province di Napoli e Caserta.

Without the following funding bodies, the Project would not have been possible, or be continuing: The Hellenic Ministry of Culture, The Cultural Olympiad 2001–2004; The European Union; The Headley Trust, UK; Hall Analytical Laboratory, Manchester, UK; The Institute of Aegean Prehistory, USA; the National Science Foundation, USA; The Holley Martlew Archaeological Foundation, UK and USA; and Davidson and Vassar Colleges, USA. I thank them all very warmly.

NOTES

[1] The catalogue numbers in the captions to the figures refer to Tzedakis and Martlew 1999. 'Substitute' vessels referred to in the captions, were the ones used in the exhibition

to illustrate types of vessel in cases where the original vessels did not survive in sufficient quantity to be restored.

[2] Evans 1964, 195, 'Sheep-bell' pottery, M.M.I, I, 189; IV,689 n. 2 (insets).

[3] Adreadaki-Vlasaki, 1987, Ομαδα νεοανακτορικων αγγειων απο το Σταυρωμενο Ρεθυμνης, *ΕΙΛΑΓΙΝΗ: Τομος τιμητικος για τον Καθηγητη Νικλαο Πλατωνα*, Herakleion 1987, 66, σημ 33.

ABBREVIATIONS

ISOB	Istituto Universitario Suor Orsola Benincasa, Naples
MM	Archaeological Museum of Mycenae
PAM	Archaeological Museum of Pireaus
PM	Archaeological Museum of Patras
PS	Pseira excavation number
RM	Archaeological Museum of Rethymnon, Crete
SM	Archaeological Museum of Sparta
X	Chrysokamino excavation number

REFERENCES

Andreadaki-Vlasaki, M., 1991-1993, Αρχαιολογικες Ειδησεις, 1989–91, *Κριτηκη Εστια*, 4, 241–244.

Andreadaki-Vlasaki, M., 1994-1996, Αρχαιολογικες Ειδησεις, 1992–94, *Κριτηκη Εστια*, 5, 251–264.

Betancourt, P. P. et.al., 1999, Research and Excavation at Chrysokamino, Crete, 1995–1998, *Hesperia*, 68, 343–370.

Betancourt, P. P. and Davaras, C. (eds.), 1995–2001, *Pseira I–V*, University of Pennsylvania Press, Philadelphia.

Evans, J., 1964, *Index to the Palace of Minos*. Biblo & Tannen, London.

Godart, L. and Tzedakis, Y., 1992, *Témoignages Archéologiques et Épigraphiques en Crète Occidentale du Néolithique au Minoen Récent IIIB (Incunabula Graeca XCIII)*. Gruppo Editoriale Internazionale, Rome.

Marazzi, M., 1998, *Missone Archaeologica Vivara (Guide to Excavations, 1994–1998)*. Istituto Universitario Suor Orsola Benincasa, Naples.

Marazzi, M. and Tusa, S., 2001, *Preistoria: Dalle cose della Sicilia alle Isole Flegree*, Istituto Universitario Suor Orsola Benincasa, Naples.

Melena, J., 1983, Olive Oil and other Sorts of Oil in the MycenAean Tablets, *Minos*, 18, 89–123.

Tzedakis, Y. and Martlew, H. (eds.), 1999, *Minoans and Mycenaeans: Flavours of Their Time*. Hellenic Ministry of Culture/Kapon, Athens, (2nd edn., 2001, Cultural Olympiad, 2001–2004).

Tzedakis, Y., Martlew, H. and Jones, M. K. (eds.), forthcoming, *Archaeology Meets Science: Biomolecular and Site Investigations in Bronze Age Greece*. Oxbow Books, Oxford.

Vassilogamvrou, A., 1996–97, Πρωτοελλαδικο νεκποταφειο στο Καλαμακι Ελαιοχωριου-Λουσικων Αχαιας, Πρακτικα Σενεδριου πελοποννησιακων Σπουδων, Αργος-Ναυπλιο, 6–10 Σεπτ. 1995/*Proceedings of the Society of Peloponnesian Studies, Argos-Nauplio, 6–10 Sept. 1995* (Πελοποννεσιακα, Supp. 22), I, 366–399. Society of Peloponnesian Studies, Athens.

Chapter 8

The Production Technology of Aegean Bronze Age Vitreous Materials

M. Panagiotaki, Y. Maniatis, D. Kavoussanaki, G. Hatton and M. S. Tite

Abstract
Contacts and technology transfers among the people of the Aegean, Egypt and the Near East are best witnessed in the surviving artefacts. This paper attempts to elucidate such contacts through the study of vitreous materials. First, the developments in the iconography and production technology of faience and Egyptian blue frit, both of which were introduced into Minoan Crete at the end of the 3rd or the beginning of the 2nd millennium BC, are discussed. The transfer of these technologies to the Mycenaean mainland, together with the first appearance of glass and new forms of faience in the Aegean, including a cobalt blue vitreous faience, are then considered. Finally, the preliminary results of the scientific examination, using analytical scanning electron microscopy, of a small selection of samples are presented.

The data presented confirm that, although the techniques of first faience and Egyptian blue frit making and later glass working were "borrowed" from the Near East and/or Egypt, the main body of the Aegean vitreous objects was made in the Aegean. Furthermore, the faience and Egyptian blue frit were produced in the Aegean using locally available raw materials whereas the glass was produced in Egypt and possibly the Near East, and imported into the Aegean as primary ingots that were worked locally.

I INTRODUCTION

Aegean vitreous materials (faience, vitreous faience, Egyptian blue frit, glass) have been the focus of our study, which involves macroscopic (M. Panagiotaki) and analytical (V. Maniatis, M. S. Tite) examination as well as experimental replication (Ch. Sklavenitis) in order to understand the methods used in their production (Panagiotaki 1995, 1999a, 2000). We should point out, however, that the study has been hindered by two major problems: one, the bad state of preservation of the material and two, the fact that well preserved objects cannot be analysed or sometimes not even handled since each is a unique piece: no mass production exists in the Aegean with the exception of the Mycenaean glass relief beads.

Therefore, macroscopic examination has been of very great importance throughout this study. In this way, it has been possible, for example, to locate the tiny specks of glaze surviving usually on the underside or edge of a faience object, or the area with a greenish tint on another, or even the glaze drip on a better preserved item. The results of the macroscopic examination are presented in sections II and III. The preliminary results of the subsequent scanning electron microscope (SEM) examination of samples taken from selected vitreous objects are presented in section IV. In addition, a start has been made on supplementing the SEM examinations using laser-induced breakdown spectroscopy (LIBS) (Anglos 2001). X-ray fluorescence spectrometry (XRF) will also be used on better preserved faience objects in the Laboratory of the National Museum in Athens (E. Mangou). The advantage of the LIBS and XRF techniques, applied on faience objects, is that they can provide qualitative or semi-quantitative analyses without the need to remove a sample, and can therefore be used in the examination of well preserved objects.

The macroscopic study has so far shown that the main body of Aegean Bronze Age faience, as well as Egyptian blue frit and glass fits into the general Aegean iconography and is thus considered as a characteristic Aegean product. There are, however, technical, compositional and even stylistic similarities with both the Near East and Egypt. Such similarities are used in this paper to elucidate contacts and exchange of technical skills and ideas, and raw materials among the ancient craftsmen.

II 3RD MILLENNIUM – EARLY 15TH CENTURY B.C.

Aegean faience first appeared in Minoan Crete in the 3rd millennium B.C. (a number of beads were also found in the north of Greece, see Mirtsou et al., 2001), Egyptian blue frit perhaps slightly later, whereas glass does not appear until the 15th century B.C. Glazed steatite or quartz stone (found in Egypt from c. 4000 B.C., see Andrews 1990, 7) have not been found in the Aegean so far (Panagiotaki hasexamined large numbers of steatite beads using LIBS) and this may be taken as evidence that the technique of faience-making was learned elsewhere and brought to Crete as a full-blown technique. It should be pointed out that all the raw materials for the production of faience can be found in the Aegean. Quartz is found everywhere, especially in Crete and the Peloponnese. Plant ash can be obtained from plants such as *Salsola kali* (rich in potash) that can be found all over the Mediterranean including Crete; *Salsola soda*, on the other hand, is not found in Crete but is found on other Aegean islands and on the mainland (Triantafyllidis 2000, 34–5). Metal ores can also be found in many places in the Aegean (Branigan 1974, 59–66). Also, Pliny refers to a source of the natural evaporite, natron, in Macedonia (Ignatiadou 2004).

II.1 Faience

Among the earliest faience objects found in Minoan Crete is a typical Pre-dynastic Egyptian vase from a burial cave in Maronia Siteias (context: c. 3000–2700 B.C., Figure 8.1): its core material is hard and well preserved; it is of cream/beige colour with a turquoise/green glaze which exhibits signs of incomplete efflorescence (Panagiotaki 2001, fig. 72). The presence of this vase as well as a large number of other Egyptian artefacts of various materials, found especially in the Mesara tombs in Crete, point to contacts and exchange of ideas between Crete and Egypt.

Large numbers of faience beads found in many Early Minoan and early Middle Minoan tombs (c. 3600–1900 B.C.) in the Mesara (Xanthoudides 1924, pls. lviii, xxvi, for Platanos and Koumasa, Alexiou 1960, for Levin) and East Crete (Money-Coutts 1935–1936, pl. 19e, fig. 26e, for Trapeza cave and Seager 1912, fig. 36, for Mochlos) are of 'common' shapes found also in Egypt and the Near East such as spherical, cylindrical or discs (for bead shapes see Beck 1928). Often they are found together with gold beads or ornaments that recall Egyptian or Near Eastern items. The Mochlos faience beads (Seager 1912, figs. 20, 25) for instance (spherical and discs) were found together with gold beads (cylindrical and other shapes) which recall Egyptian beads (Andrews 1990, fig. 62); the gold ornaments (other than beads) from the same site recall examples from the Royal cemetery at Ur (Frankfort 1996, fig. 68) suggesting perhaps contacts with both Egypt and the Near East.

Figure 8.1 Faience Egyptian cup from a burial cave in east Crete.

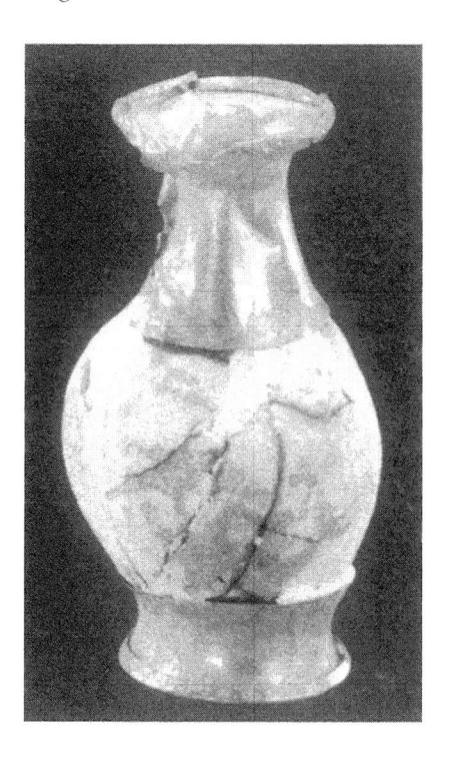

Figure 8.2 Faience miniature vase trimmed with gold from the Palace of Knossos.

All the Early Minoan faience beads consist of a white, or dark brown fine-textured core material and traces of a bluish or mostly greenish glaze, which is now matte and badly weathered. Both the cylindrical beads and discs were made round a wire/plant stem, and then sliced, while the spherical ones were made as a ball, which was pierced when the material was still soft, the pressure exercised often distorting the shape of the bead. A few cylindrical beads may have been glazed by the application method since their perforations are unglazed and the glaze around the perforations looks as if it was touched when it was wet (suggesting that the beads were touching one another as they were drying).

The Middle Minoan I–II period (2160/1979–1700/1650 BC) is again represented by beads and a few inlays. The (MM IB) large annular beads from the Vat Room Deposit at the Palace of Knossos are of white core with traces of a green glaze (Evans 1921, fig. 120 centre; Panagiotaki 1999b, 8e); small inlays from the same deposit are of the same texture and colour (Panagiotaki 1999b, 8d). The (MM II) Quartier Mu beads at Malia are rectangular of brown core with a purple glaze and decorated with circles in relief; they are unique in Crete until now and they do not seem to have any parallels anywhere else (Detourney et al. 1980, 133–4, fig. 186). Unique is also a (MM II) miniature vase (Figure 8.2) made of faience and gold (Evans 1921, 252, fig. 189a; Foster 1979, fig. 1; Panagiotaki 2001, fig. 73), a combination that recalls vases from Egypt and the Near East (Lilyquist and Brill 1995, fig. 10).

Figure 8.3 Faience House Façade from the Palace of Knossos. Courtesy of the Ashmolean Museum.

Although it is not entirely clear if these early faience objects were made in Crete or elsewhere, the fact that their core material and glaze do not look different from the later faience from the Palace of Knossos (New Palace Period), where there must have been the major faience production center of Minoan Crete, points rather to their being made in Crete (see also section IV); Foster (1987b) had suggested the existence of a workshop immediately west of the Palace, but this needs more investigation.

During the Middle Minoan IIIA period (1700/1650–1640/1630 BC) the best known faience objects come from the Palace of Knossos: the Town Mosaic House Facades and related plaques (Evans 1921, figs. 223–226, 228–230; Foster 1979, 99–115; Panagiotaki 2000, 155). They depict house facades (Figure 8.3) but also human and animal figures and plants. The faience core material exhibits a great variety in colour: different shades of green, grey and brown but also white and red. The plaques are made in moulds; our experimental replication showed that the tiny knobs and the minute inlaid work on them could not have been achieved otherwise (Panagiotaki *et al.* in press). The human figures are formed separately and stuck onto the flat plaque, creating thus, low relief work. A lot of the plaques are inlaid: the plaque came out of the mould with the design in the form of shallow channels, which were subsequently filled with a paste of fine core material of contrasting colour. White paste is inlaid in a grey or brown core; grey or brown paste in a white core. The glaze survives only as little patches; it is white or green, but other colours too may have originally adorned these plaques. It may have been achieved by the application method since efflorescence or cementation would not have worked because of the minute inlaid work. Inlaid work (and layering) appeared first in Egypt but nowhere was it developed the way it was in Minoan Crete (Kaczmarczyk 1983, 305; also Vandiver (1983, 93); see also Lacovara 1998, figs. 20, 21–3 for the Kerma inlaid faience of 1600 B.C.). The

Minoan inlaid pieces have quite intricate designs and colour combinations and the inlaid paste fits so perfectly that it gives the impression of paint work and not of inlays (for the colour compositions of Minoan faience see Foster 1987a). Inlays are difficult to achieve as the paste to be inlaid has to go into the prepared channel at the right moment, when the body is neither too dry nor too wet, otherwise the inlay would shrink away. The separation of the inlay from the body, which becomes fashionable in New Kingdom Egypt, is never seen in the Aegean (Vandiver 1982a, 176).

The Town Mosaic House facades recall the concept behind the tiles that decorated the pyramid of the IIIrd Dynasty King Djoser in Egypt (Lauer 1936, 36–8; Friedman 1998, nos. 17–20). However, the Egyptian plaques were considerably larger and also monochrome (green). The Knossos plaques are polychrome and apparently Aegean in their iconography: they present not only house facades but also animals in their natural settings and human beings. A whole town may have been presented, perhaps with animals grazing, people perhaps swimming in the sea and warriors. All is executed to minute scale, recalling the later miniature frescoes from the Palace of Knossos (Evans 1930, 31–65) as well as the town fresco at the West House on Thera (Doumas 1992, 15).

The decorative objects from the 6th Kasella in the West Magazines were enveloped with gold leaf (Evans 1921, 324). Their core material is of fine texture in brown and the glaze is of a purplish tint; however, closer examination made it clear that there were small patches of blue/green glaze (the most usual glaze colour in Aegean faience, created by the use of copper oxide). The objects were involved in a fire, which melted part of the gold leaf, which can now be seen as tiny granules of varying size on the surface of the objects. It is not clear if the glaze became purple as a result of the same fire alone: when Panagiotaki submitted replicas of faience glazed with copper oxide in an open fire the blue/green glaze became black in places, not purple. LIBS analyses applied on the above decorative objects showed a lot of manganese in the body and some in the glaze, besides the copper that created the green glaze; it is thus possible that the purple colour was the result of the combination of copper and manganese.

In the Middle Minoan IIIB/Late Minoan IA (1600 BC) or Late Minoan IA (1600/1580–1480 BC), the Minoan faience-maker is a skillful painter and sculptor. Many objects have been decorated with black linear designs and then glazed (Panagiotaki 1999b, pls. 9a, 10c). A series of plaques with inlaid bands (Evans 1921, fig. 344) from the Palace of Knossos continues the Middle Minoan IIIA tradition, they are inlaid creating alternate bands: white combined with brown or grey but not to the minute scale, so characteristic of the previous period. Plaques in various shapes inlaid with paste of contrasting colours or painted with contrasting colours, knobs and buttons, arrow shaped items or round tablets were used to decorate furniture and walls; some have dowels on their underside to fit into other plaques. There is a tendency to use a white or cream faience core with brilliant green or turquoise glaze, or a brown or grey core in most cases inlaid with white paste and again a green glaze over it. The Throne Room round tablets, for instance, are of dark brown core, inlaid with a white paste in the shape of a lozenge and the white lozenge is further inlaid with very fine brown stripes. Replication of inlaid plaques (white paste inlaid into brown core material) showed that the glaze can be achieved by the application method and has to be a very thin coat, otherwise the white inlaid bands do not show under the glaze.

A fruit, two flowers and a cockle shell from the Palace of Knossos have two different layers of glaze (Panagiotaki 1999b, pl. 9d, e): an all-over turquoise and a brown glaze only on the upper surface; in these cases the turquoise could have been made by the efflorescence, the application or the cementation method, while the second layer could have been achieved only by application – brush marks are often visible along the edges of the objects.

In the few cases where the core material was not of very fine quality, a fine layer of white material was used (what Lucas called Variant A, see Lucas and Harris 1962, 141). It was mostly applied on larger objects such as the dress plaques from the Palace of Knossos (Panagiotaki 1999b, fig. 27).

Hundreds of beads were made in this period too; they are mostly spherical or in the shape of the grain of wheat. An unusual, and unique in the Aegean, bead shape, however, comes from the Mesara, from the MM III–LM I Kamilari Tholos tomb (Figure 8.4, see Levi 1961–1962 fig. iv; Panagiotaki 2001, fig. 87b) and may have originally come from Egypt (or the idea came from Egypt), where there are some examples (for a later in date necklace (1370–1320 BC) see Andrews 1990, title page, second from top): the beads are in the form of a 'four-celled fruit' or four granules of the same size joined together (on granulated beads see Lilyquist 1993); they are made of white core with purple glaze. A bead from the Palace of Knossos (Evans 1921, fig. 356) is in fact a unique pendant of a 'waz lily' in relief (created in a mould), which can be seen as the forerunner of the gold and glass relief beads that were to dominate the whole Aegean world in the next period.

Some vases were made in parts, in moulds, and the pieces were subsequently joined together in a way that they support each other. A tall cup from Knossos was made in two pieces, the cylindrical body first, while the wide rim was positioned in such a way as to be supported by the upper part of the body (Evans 1921, fig. 357; Panagiotaki 1999b, 15d,e). Decorative elements (a rose branch) are

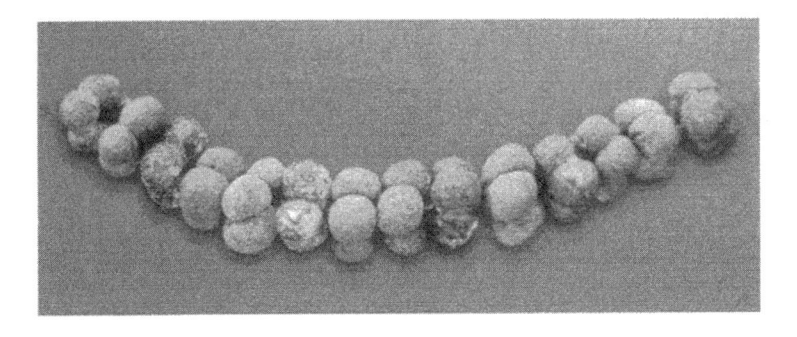

Figure 8.4 Faience granulated beads from south Crete.

made separately and stuck on or they are built up from successive layers of a thick paste (plants). The inside of the vase is partly covered in glaze (the lowest part is free of it), drips suggest that the vase was immersed upside-down in a liquid glaze (the method of application) and then stood upright to dry so that the liquid glaze created a kind of a rim with runnels and drips on the inside of the vase.

A series of plaques with animals in low relief, the Cretan *agrimi* or cows and bulls from the Palace of Knossos were probably made in a mould for the basic shape, but many body details were subsequently ground on, when the faience was not too dry, but certainly before firing (Panagiotaki 1999b, figs. 17–8, 20, pls. 12, 13 and 21). Body details are picked out in black paint, usually placed on the faience body before glazing. The animal plaques show, that the faience-maker continued to create syntheses recalling the Town Mosaic but to a larger scale and displaying a closer adherence to the natural world. The plaques may have formed

Figure 8.5 Faience snake figure from the Palace of Knossos.

not a town as in the case of the Town Mosaic but a bucolic scene. The whole may have been set onto a wall to counterbalance the colourful Minoan frescoes; the idea again recalls the monochrome tiles covering the walls in the Pyramid of King Djoser or the colourful tiles representing captives, gaily dressed, in a much later palace, that of Ramese III (c. 1194–1163 BC – Crowell 1998, nos. 52–4) but the setting and the execution is Minoan, and so is the treatment of the animals which betray a powerful hand capable of bringing out the characteristics of such animals, like the alert look of the Cretan *agrimi* or the slow movement of a cow, using a material whose lack of plasticity was not helpful.

The human figures in the round from the Temple Repositories at the Palace of Knossos (Figure 8.5) prove the skill of the Minoan faience-maker as a sculptor (Evans 1921, 359–60). They are made in pieces and then joined together. Details are stuck on, such as the eyes, eyebrows, locks of hair and snakes. The tight bodice gives the impression of a fine, almost transparent material, which has been embroidered and is trimmed with black straps. The flounced skirt and the raised arms of the younger figure give the impression that she moves the same way dancing figures seem to move in frescoes from the Palace.

In this period faience penetrates the Mycenaean world (mainly the Peloponnese). The Shaft graves yielded faience which seems contemporary and of the same manufacture as many faience objects from Minoan Crete, especially from the Temple Repositories (on the date of the Shaft Graves see Warren and Hankey 1989, 96 with the relevant bibliography). A tall cup from Shaft Grave A of Grave Circle B (Foster 1979, 123, fig. 32) is almost identical with the two cups from the Temple Repositories (see above) and was perhaps made by the same artisan or school. The Ostrich eggs rhyta from Shaft Graves IV and V as well as from Thera have mouthpieces and decorative appliqués that again recall Minoan faience (Karo 1930, 114ff. figs. cxli-cxlii; see also Sakellarakis 1990, 285–307). Two pairs of 'sacral knots' come from Shaft Grave IV at Mycenae (Karo 1930, fig. cli; Foster 1979, 140–1, pl. 46): they show a long scarf-like piece of material bound to create a loop, a knot and the two ends of the 'scarf' which terminate to a tassel fringe. The cloth depicts a tartan design in brown and white; one sacral knot has a brown background with very fine white stripes, the other a white background with brown stripes. The sacral knots were made in pieces and stuck together; their underside is mostly hollow and shows the way the material was manipulated. Variant A is used to cover the beige core material, while the stripes that create the tartan design are very fine channels filled with fine white or brown paste. Traces of a green glaze survive in places (only on the upper surface and edges), which could have been achieved only by the method of application because of the inlaid work involved. From the same grave also come large rosettes combined with a lozenge to create part of a gaming board (Karo 1930, fig. clii:556), analogous to the board from the Palace of Knossos (Evans 1921, pl.v).

These 'Mycenaean' objects seem to be if not of Minoan workmanship at least made under Minoan influence. A vase (NM 7509) in the shape of a lion's head from the Acropolis at Mycenae is among the few objects that may not have been Minoan work. The core is of off-white colour and is covered with a green glaze;

the mane of the lion is created by shallow grooves and round depressions filled with black slurry (that seems to have been applied over the green glaze), which spreads and covers most of the head – a rather unusual treatment by Minoan aesthetic standards.

II.2 Egyptian blue frit

Egyptian blue frit was first made in Egypt during the Old Kingdom (2613–2181 BC) (Lilyquist and Brill 1995, 5). Its first occurrence in Crete is in the Vat Room Deposit at Knossos (19th cent BC, Figure 8.6) where almost two thousand minute beads of the type made round a wire as a long rod and then sliced to irregular sizes were found (Evans 1921, 170; Panagiotaki 1999b, 40–1; see also section IV, 173–4).

Egyptian blue frit beads were also found in MM and LM tombs, mainly in shapes such as cylinders, grains, discs, papyrus heads, spherical and melons (Lembesi 1967, 195–209; Muhly 1992, 92–93, 129, fig. 27; Panagiotaki 2001, fig. 98). At the Palace of Knossos it was employed to colour the crystal lozenges in the royal gaming board (Evans 1921, pl. v) as well as a rock crystal plaque from the Throne Room, where a bull is depicted in full charge (Evans 1930, pl. xix), but also in frescoes (Cameron 1987, 321–28).

III LATE 15TH – POST 15TH CENTURY

III. 1 Faience

During the Late Minoan period (LM IB 1480–1425 BC) large pieces of faience sculpture are produced in Crete. The faience-maker continues to draw his inspiration from the animal world: a large Argonaut (Figure 8.7) rendered in a very naturalistic way and the rhyta in the form of bulls' and lions' (or wild cats') heads from the Palace of Zakro (Platon 1974, fig. 88) seem to have been made in moulds, perhaps in different pieces. A triton shell from the country house at Myrtos Pyrgos (Cadogan 1981; 2000, fig. 70), in East Crete, was also made in a mould and so was another from Shaft Grave III at Mycenae (Karo 1930, fig. cxlviii:166; Foster 1979, 137–8, pl. 44). These large pieces of faience as well as a vase with double almost flaring mouth also from Shaft Grave III at Mycenae (Karo 1930, 242, fig. cxlix:166; for other vases fig. cxlviii) consist of a fine textured core material in white, with an often pink tint and a green glaze, which has reddish patches over it; this combination of colours, which is characteristic of this period, is now under scrutiny. Since the majority of these vases come from the Palace of Zakro, it is possible that they were all made there (on workshops at Zakro see Platon 1974, 200 and L. Platon 1993).

The faience vases from the House of Shields at Mycenae (LM IIIB 14th – 12th century) present a different case (Wace 1956, 107–113; Foster 1979, figs. 35–40; Andreopoulou-Mangou 1988, fig. 12; Peltenburg 1991, 165–6; Tournavitou 1995); they are large with linear decoration or they are made of polychrome faience, the glaze of which is well preserved and glossy. In a strap handle of a vase (NM

Figure 8.6 Egyptian blue frit beads from the Palace of Knossos.

Figure 8.7 Faience Argonaut from the Palace of Zakros.

7506) and in a few fragments of another (showing the heads of a lion and griffin) the glaze is rich and glossy and with colour combinations (blue, yellow, red) that recall some Amarna faience glazes (Leveque 1998 fig. 196). We achieved replicas with such rich and glossy glaze when we applied glass (the Mavro Spilio glass –

Figure 8.8 Faience beads from tombs in the Peloponnese.

according to our analysis) over unglazed faience objects. A rhyton (Foster 1979, 134, pl. 43 – NM 2625, see also section iv) decorated with spirals bordered by cables in a soft brown, in an all-over white glaze (white glaze is found only in two or three cases in the whole Aegean), recalls the white glaze of a perfume vase from Nubia (Friedman 1998, fig. 82) of the 18th Dynasty while the spiral decoration recalls the Kition rhyton from Cyprus (Peltenburg 1972; 1974). These few Aegean vases with a glossy glaze which is close to the Egyptian thick glossy glaze of some Amarna pieces (Friedman 1998, fig. 25; Crowel 1998, fig. 33) may have been made in an Aegean workshop following an Egyptian recipe for the glaze, **or** they were made elsewhere but for the Aegean, if we judge by the shape and subject matter, both of which match Aegean vases (see also Panagiotaki 2004, and forthcoming).

During the late 15th, but especially during the 14th and 13th centuries a great variety of faience beads were made (rosettes, papyrus heads, spherical, grain of

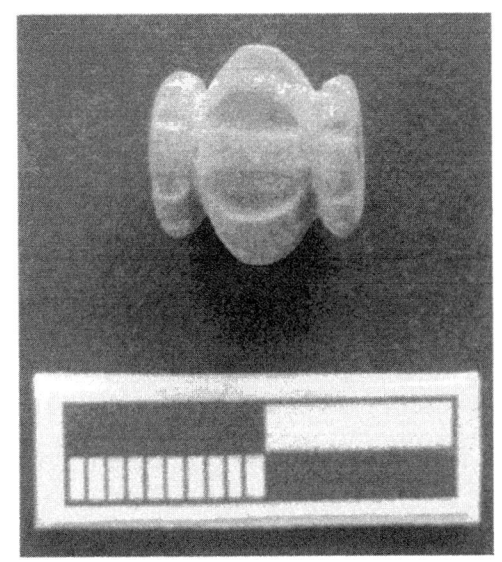

Figure 8.9 (left) Faience bead from tomb in Crete.
Figure 8.10 (right) Vitreous faience bead from Attica.

wheat, flat amygdaloid, flattened spherical with grooves, spherical collared with grooves, melons). However, two, new to the Aegean, kinds of faience can now be identified (both in bead form). The first, used to produce flat amygdaloid beads (Figure 8.8) as well as the flattened spherical beads with grooves (Figure 8.9), consists of a yellow/orange core material with traces of a grey, matt glaze that is all that remains from a dark blue glaze (cobalt blue), now surviving only as rare specks or patches. The second, used to produce collared beads with grooves (Figure 8.10) and grain shaped beads, is a vitreous faience coloured grey throughout. The few examples of vitreous faience from Crete do not retain any traces of glaze. However, some of the much larger number of vitreous faience beads from the mainland, coloured the same grey and of the same shape, retain specks of a brilliant, dark blue glaze (cobalt blue again). Most of the collared with grooves beads come from the Tholos tomb at Menidi in Attica (14th–13th century) (Lolling 1880; Demakopoulou 1998, fig. 59), while others were found at Prosymna (Blegen 1937, fig. 285:7) and at Mycenae (Xenaki-Sakellariou 1985, tomb 91) in the Peloponnese, while a few more were found in Crete in tombs in Katsamba (Alexiou 1967, fig. 35, 37) and at Gournes. Amygdaloid beads are found in both Crete (at Gournes and Gournia, unpublished) and the Mainland (at Asine, see Frödin 1938, fig. 266, at Prosymna, see Blegen 1937, fig. 285, at Mycenae, see Xenaki-Sakellariou 1985, tombs 55 and 58). The flattened spherical with grooves are found around the Knossos area (at Zafer Papoura, see Evans 1906, fig. 81a) but also in the east part of the island (at Kalou, Mouliana Siteias, unpublished) as well as in tombs in the mainland (at Aidonia cemetery, see Demakopoulou 1998, fig. 57, at Prosymna, see Blegen 1937, fig. 199:2, at Asine, see Frödin et al. 1938, fig. 266).

Some bead shapes found in the Aegean are also found in the Near East and Egypt but it is not clear whether they all had a common source or simply the shapes were imitated. Such shapes are: the flattened spherical with grooves also found at Tell Brak (Oates et al. 1997, fig. 251:6–66) in Mesopotamia and in Italy (Bellintani 2003, Fig. 1.5), the daisy-like from Gournia (unpublished) also found at Tell Brak in blue frit (Oates et al. Fig. 241:44), the granulated also found at Nuzi in Mesopotamia in red frit (Vandiver 1982b, fig. 4s) and in the Levant in faience (Spaer 2001, fig. 27b), the lanteen-shaped (Foster 1979, Fig. 97) also found at Tell Brak (Oates *et al.* Fig. 251.67) and in Italy (Bellintani 2003, Fig. 1.4), and finally the corn flower shaped or in the form of a vase were commonly found in Egypt (Andrews 1990, figs. 27 and 158 for gold examples).

III. 2 Egyptian blue frit

Small vases were produced in this period but they survive only in tiny fragments and are mostly unpublished. Vase fragments, pot stands as well as lumps of the coarser raw material have been retrieved at Knossos (Cadogan 1976, 18; Cline 1994, 221); more lumps were also found in a tomb in the Mesara (Panagiotaki 2000, 157). It is, however, much later, in the first millennium (7th–6th cent B.C.) that blue frit is used to create large and smaller vases decorated with incised designs or basket-work (Panagiotaki 2000, 158; 2001, fig. 381).

Beads made of blue frit continued to be produced, in fact numbers increased in the mainland and the simple shapes of the last period became more complex: papyrus heads, rosettes of various sizes and shapes, cockle shells (Xenaki-Sakellariou 1985). An unusual shape bead comes from the cemetery at Aidonia in the Peloponnese and is in the form of a duck (Demakopoulou 1998, fig. 33:a) with its head turned back and stuck onto its neck (unlike the more common Aegean type where the duck keeps its head up).

Egyptian blue frit is also used in frescoes at the Palace of Knossos and the House of the Frescoes (Evans 1921, figs. 397–8; on the dates of some frescoes see Hood 1978) and decorated the Temple Tomb at Knossos (Evans 1935, 974–5) and tomb H at Katsambas (Alexiou 1967, 35).

III.3 Glass

Sometime in the 15th century glass was introduced to the Aegean – a more exact date cannot be given since most glass objects come from tombs which were used for long periods of time; glass objects, recently identified by the author among material excavated at the Palace of Knossos by Sir A. Evans, do not come from datable contexts either. The only more exactly dated glass comes from a building west of the Palace at Knossos (Cadogan 1976, 18–19), dated to LM IB (1480–1425), and has been suggested that glass was worked there. The idea that glass was worked at Knossos is further supported by a few amorphous pieces of dark blue glass, which may have come from another working area at or by the Palace. More evidence of glass working comes from Tiryns in the Peloponnese from where half finished objects, rods and drops of glass were recently identified by the author (Panagiotaki 2004). The idea of glass-working at Knossos is further supported by a series of glass objects, previously published as of faience (Evans 1928, fig. 440; Panagiotaki 1999a, 621, fig. cxxix:c), which are made with known faience objects as their prototype. Glass figures from the palace of Knossos (with a flounced skirt Fig. 8.11) seem to continue the Minoan faience tradition but the uncertainty of the crafts person with the new material is obvious as the figures are not free standing like their faience counterparts, but plaques, to lie flat as inlays.

Glass vases found in Crete and in the mainland do not look stylistically different from Egyptian vases and a few have been considered to be Egyptian imports. For example HM 199 of grey to bluish glass with white and yellow

Figure 8.11 Glass figurine from the Palace of Knossos.

decoration, of Dynasty XVIII (Phillips 1991, no. 77) and another vase from Karteros (Weinberg 1961–1962, 226–229, fig. mh 1; Philips 1991, no. 87). They mostly carry white and yellow thread decoration typical of Egyptian vases (Hayes 1990, vol. i, fig. 109); exceptions are: the shape and decoration of HM 270, which recalls a faience vase from Cyprus (Peltenburg 1974, pl. lxv), and another from Kalyvia (15th – 14th cent), which is core formed and decorated with a star in white (Savignoni 1904, 556–557.). The only vase that has been seen as an Aegean product is a vase from Kakovatos in the Peloponnese made in a mould (Chatzi-Spiliopoulou 2002, 69, fig. 5).

Clearly imported glass objects perhaps from Mesopotamia as diplomatic or bridal gifts include figurines (also worn as pendants) of the fertility goddess Ishtar (one was found at the Tholos tomb at Kakovatos see Müller 1909, 278, fig. Xii.5:6; and Chatzi-Spiliopoulou 2002, 67) and also discoid pendants depicting a rayed star with spheres between the rays also associated with the goddess Ishtar in the Near East and Egypt, which were found on the Akropolis at Mycenae, at Thorikos in Attica, the Tholos tombs at Kakovatos and Dara in the Peloponnese (Chatzi-Spiliopoulou 2002, 67–8).

What is, however, immediately recognized as a characteristic Aegean glass object, is the so-called Mycenaean glass relief bead and decorative plaque found in every tomb in the Aegean world (on the shapes and types of Mycenaean glass relief beads and plaques see Chatzi-Spiliopoulou 2002, 63–87; on the use and symbolism of beads see Hughes-Brock 1999, 277–96). A great variety of shapes exist, always drawn from the natural world, especially flowers, plants and sea creatures. The most common motifs are rosettes (single or double), ivy leaves, seashells (especially cockles and trochuses) and stylized octopuses.

The beads are mostly in dark blue, in a few cases almost transparent (or at least letting light pass through), and in one case (involving 4 or 5 plaques) in almost black (matt) (NM 7366, from Dendra, see Persson 1931); others are in translucent turquoise (NM 7369, from Dendra). Glass beads and plaques found in tombs in the mainland are better preserved than those from tombs in Crete.

Glass relief beads were made perhaps in stone moulds as a softened glass pressed or molten glass poured into the mould (on the use of moulds see Grose 1989, 31; Stern and Schlick-Nolte 1994, 49–50; Tournavitou 1997, 211–30, 243–53; Chatzi-Spiliopoulou 2002, 70–77), or even as powdered glass that was melted in the open face mould itself (we are now in the process of experimenting in making glass using various types of moulds). After cooling the object was trimmed and often abraded to acquire the desired shape – working marks are often obvious in the finished product. The beads and often the plaques were decorated with gold leaf or in some cases the whole bead was entirely covered with gold leaf (Frödin et al. 1938, fig. 102:a; see also Xenaki-Sakellariou 1985).

The glass relief beads and decorative plaques exhibit great artistic skill that obviously suited the aesthetic demands of the Mycenaean Aegean as well as the beliefs and notions attached to them. Judging from the sheer numbers of glass beads and plaques found, it is certain that glass caused the dwindling in faience but it may have given an impetus to Egyptian blue frit, vitreous faience as well as to yellow/orange ordinary faience with dark blue glaze, both of which appeared at the same time or soon after the introduction of glass.

IV CONTRIBUTION OF SCIENTIFIC EXAMINATION

The primary aim of the scientific examination was to obtain information on the sources of the quartz, alkali and colorants used in the production of the faience, Egyptian blue and glass objects found in the Aegean and, in particular, establish the extent to which the Egyptian blue frit and glass were imported as "raw materials" for object production from Egypt or the Near East.

IV.1 Experimental methods

The principal technique employed in the scientific study of the Aegean vitreous materials was the examination of polished sections in analytical scanning electron microscopes (SEM). The microstructures of the samples were observed in backscatter mode in which the different phases present could be distinguished on the basis of their atomic number contrast (i.e. quartz particles appear dark as compared to the higher atomic number calcium/copper-rich phases that appear light). Energy dispersive spectrometry (EDS) was used to determine the bulk chemical compositions of the different vitreous materials and, when available, wavelength dispersive spectrometry (WDS) was used to determine the compositions of the individual phases present.

The major problem with the scientific examination of vitreous materials from the Aegean is that a high proportion of the material is severely weathered so that minimal glass phase survives, the problem increasing with increasing porosity of the material. Thus, the problem is greatest with faience and Egyptian blue frit. The glaze and interstitial glass of the former and the glass phase of the latter, even when surviving, are usually seriously depleted in alkalis. For the less porous vitreous faience, the problem is less severe with interstitial glass, in which the alkalis survive, frequently being found away from the surfaces of the objects. Similarly, an unweathered core frequently survives in all but the thinnest glass objects.

Preliminary results of the SEM examination for faience, Egyptian blue frit and vitreous faience (considered separately from standard faience) are discussed below together with published analytical data for Aegean glass (Brill 1999). In trying to establish whether the Egyptian blue frit and glass were produced locally in the Aegean or imported as "raw materials" from Egypt or the Near East, the principal comparison is with analytical data from a detailed study of vitreous materials from the New Kingdom site of Amarna in Middle Egypt that was occupied during the mid–late 14th century BC (Shortland 2000, Tite and Shortland 2003). Although the Amarna material is later in date than most of the material considered in the present study, the associated analytical data is representative of vitreous materials produced in Egypt and the Near East from the beginnings of glass production at about 1500 BC.

IV.2 Faience

The two possible sources of quartz for the faience bodies are either crushed quartz pebbles or quartz sand. Because of the very fine particle size of the quartz in the majority of Minoan faience bodies (Figure 8.12), if sand was used it would almost certainly have had to have been subjected to further grinding. Therefore, the criterion that angular quartz particles indicate crushed quartz pebbles, rather than quartz sand, is not applicable. However, the fact that the alumina contents of the faience bodies are typically less than 1% suggests that crushed quartz pebbles rather than quartz sand were used (Brill 1999).

The two possible sources of alkali used in the production of the faience glaze are plant ash and natron. The plant ashes can be either soda or potash rich, depending on the type of plant used (Brill 1999). Additionally, plant ashes contain small amounts of lime and magnesia (Sayre and Smith 1974). In contrast, the natural evaporite, natron, consists predominantly of sodium carbonate and sodium bicarbonate and contains very few impurities. Unfortunately, as a result of weathering, the data on the alkalis used in Aegean faience are minimal. No unaltered faience glaze has been found (Figure 8.12). Even the glaze on the Mycenaean rhyton (National Museum 2625) which is glossy white in appearance contains negligible alkali. Similarly, unaltered interstitial glass was located only in the brown inlay of one of the Town Mosaic house façade fragments. This glass phase was shown to contain almost equal amounts of potash (5.1% K_2O) and soda (4.3% Na_2O), indicating the use of a more potash rich plant ash than the soda rich desert and coastal plant ashes used in Egypt and the Near East.

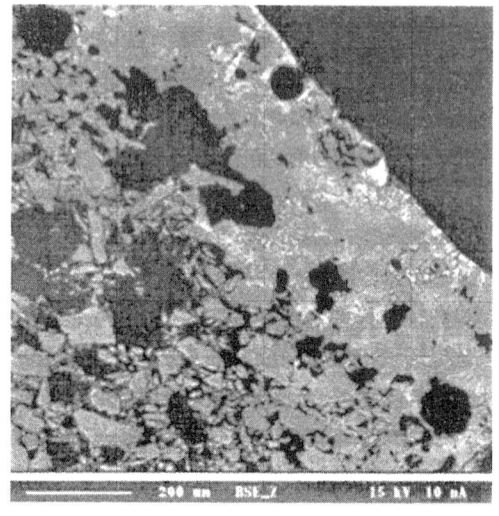

Figure 8.12 SEM photomicrographs of cross-sections through Minoan faience showing weathered glaze layer (upper right) and quartz body in which no interstitial glass is visible.

Copper oxide, added either as malachite or some other copper rich mineral, or as scale from copper or bronze metal, is present at concentrations up to few percent in all the glazes analysed, whether green, brown or purple in colour, with possible exception of the Mycenaean rhyton (National Museum 2625). The green glazes contain no other colorant. However, the brown glazes also contain a few percent of manganese oxide, some of which is present as aggregates of fine particles. Similarly, the brown bodies are coloured by manganese oxide some of which again survives as aggregates of fine particles. No measurable amounts of barium oxide were detected in any of these glazes or bodies suggesting strongly that the manganese mineral used was pyrolusite and not psilomelane (Foster and Kaczmarczyk 1982, Foster 1987a). Although the very pale green observed in the case of the copper coloured glaze on white bodies would certainly originally have been a much more intense greenish-blue, the original colour of the copper plus manganese coloured glaze on the brown or white bodies can only be rediscovered by replication.

Because of the severity of the weathering, interstitial glass, even if originally present, does not survive in Aegean faience bodies (Figure 8.12). Therefore, the microstructures observed in the SEM and, in particular, the presence or absence of interstitial glass cannot be used to identify the method of glazing used (Tite and Bimson 1986). Instead, as discussed in sections II.1 and III.1, macroscopic features such as drips or runs of glaze (Vandiver 1998), together with an assessment of the practical feasibility of obtaining the desired result, must be used to distinguish between the three possible methods of glazing (i.e. efflorescence, application and cementation).

IV.3 *Egyptian blue frit*

Egyptian blue frit is produced from a mixture of quartz, calcium carbonate and copper oxide to which is added a few percent of alkali (Tite et al 1984). When this mixture is fired to about 950°C, the alkali reacts with the quartz and lime to form a glass phase whose presence facilitates the formation of the Egyptian blue crystals (ie calcium-copper tetrasilicate – $CaCuSi_4O_{10}$) that are the source of the blue colour. The microstructure of the resulting frit, as seen in the SEM, consists of a mixture of Egyptian blue crystals and unreacted or partially reacted quartz particles, bonded together by varying amounts of a glass phase (Figure 8.13). This frit is then ground to a fine powder that can be used directly as a pigment or can be moulded to form small objects that are refired, again to about 950°C.

There are unlikely to be significant differences in the microstructures of Egyptian blue frit produced in Egypt or the Near East and that produced in the Aegean. Therefore, the main hope for establishing whether the frit was produced locally or was imported as a "raw material" from Egypt or the Near East is by comparison of type of alkali used in its production.

Unfortunately, as in the case of the Aegean faience, the glass phase originally present in Aegean Egyptian blue frit has now normally been lost as a result of weathering. The exception is the earliest Egyptian blue object from Crete to be analysed, one of the two thousand or so beads found in the Vat Room Deposit

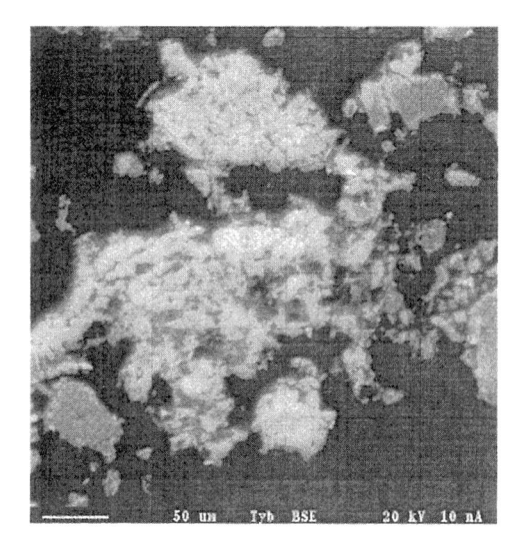

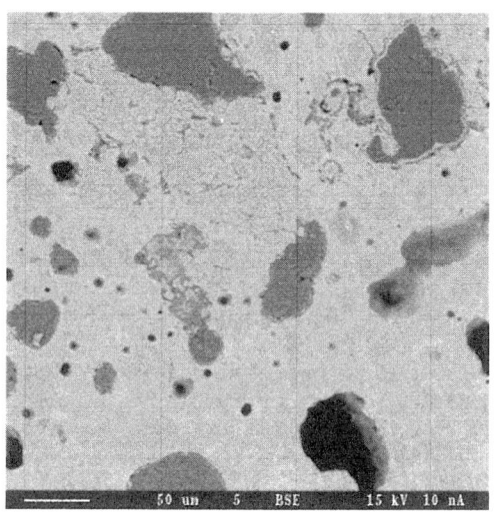

Figure 8.13 (left) SEM photomicrograph of cross-section through Minoan Egyptian blue frit showing Egyptian blue crystals (white), unreacted quartz particles (dark grey) and weathered glass phase bonding together aggregates of Egyptian blue crystals.

Figure 8.14 (right) SEM photomicrograph of cross-section through Aegean vitreous faience showing quartz particles (dark grey) in a more-or-less continuous glass matrix (light grey).

and dating to the 19th century BC. In this case, the surviving glass phase contains significantly higher potash (7.2% K_2O) than soda (4.6% Na_2O), indicating the use of a potash rich plant ash. In addition, the Vat Room Deposit Egyptian blue bead is unique in containing a scatter of fibrous calcium phosphate crystals. Therefore, it seems very probable that these early beads were again produced locally using a potash rich plant ash, fairly similar in composition to that used for the brown inlay in the Town Mosaic faience, rather than being made from Egyptian blue frit imported from Egypt or the Near East in which a soda rich plant ash would have been used.

IV.4 Glass

The cobalt blue glasses from Mycenae analysed by Brill (1999) are characterised by the high alumina (average 2.2% Al_2O_3) concentrations that are also found in cobalt blue glass from Egypt (average 2.5% Al_2O_3) (Shortland 2000, Tite and Shortland 2003) and indicate the use of cobalt rich alum from the Dakhla or Kharga Oases in the Western Desert of Egypt as the source of the cobalt colorant (Kaczmarczyk 1986). These Mycenaean cobalt blue glasses also exhibit the lower normalised potash contents (average 1.3% K_2O), as compared to those for copper blue glasses (average 2.3% K_2O), that further characterise cobalt blue glass from Egypt (Tite and Shortland 2003). Similarly, the copper blue glasses from Mycenae, also analysed by Brill (1999), are comparable in composition to those produced in both Egypt and the Near East.

Therefore, although the great majority of the glass relief beads and some of the glass objects from the Aegean were produced locally, the "raw glass" from which they were made was most probably imported into the Aegean from Egypt for the cobalt blue glass, and from either Egypt or the Near East for the copper blue glass. Further evidence for the trade in "raw glass" is provided by the cobalt and copper blue glass ingots of comparable compositions that were found on the Uluburun shipwreck (Bass 1991).

IV.5 Vitreous faience

The vitreous faience beads found throughout the Aegean in the period post 1500 BC are similar in appearance to the contemporary cobalt blue vitreous faience that was produced in Egypt in association with the beginnings of glass production, both being coloured bluish-grey throughout (Shortland 2000, Tite and Shortland 2003). The microstructure of the Aegean vitreous faience, as seen in the SEM, consists of fine quartz particles (typically less than about 60 μ across) in a more-or-less continuous glass matrix (Figure 8.14) and is therefore similar to that of the vitreous faience produced in Egypt. Furthermore, the Aegean and Egyptian vitreous faience are both coloured by cobalt and the alkalis used for both are soda rich (average 11.6Na_2O% Na_2O and 0.9% K_2O in the glass phase of Aegean vitreous faience). However, in spite of these obvious similarities, preliminary analyses suggest that these two forms of vitreous faience differ in the nature of the colorants used in their production.

Perhaps most important, the alumina contents are significantly lower in the Aegean vitreous faience (average 0.8% Al_2O_3 in the glass phase) than in the Egyptian vitreous faience (average 5.3% Al_2O_3). It therefore seems very unlikely that the source of the cobalt was the cobalt-rich alum from the Dakhla or Kharga Oases that was used in Egypt in the production of both glass and vitreous faience. As a result, cobalt blue glass that, as discussed above, was most probably imported as a "raw material" into the Aegean from Egypt could not have been used as the source of the cobalt in the Aegean vitreous faience. In addition, the Aegean vitreous faience differs from the Egyptian vitreous faience in containing, as well as the cobalt colorant (~1.5% CoO in the glass phase), comparable amounts of copper and lead (1.5–3% CuO and PbO).

Therefore, the Aegean vitreous faience was most probably produced locally using a different cobalt colorant to that used for Egyptian cobalt blue vitreous faience and glass. However, in contrast to the Minoan faience and Egyptian blue frit that were also produced locally, either a soda rich plant ash or an evaporite such as natron, and not a potash rich plant ash, was used in the production of the Aegean vitreous faience.

V. CONCLUSIONS

To conclude, although the techniques of first faience-making and later glass-working were 'borrowed' from the Near East and/or Egypt, the main body of the Aegean vitreous objects was made in the Aegean since they clearly match the Aegean iconography in which nature is glorified in the form of animals and sea creatures or flowers and plants. Further, together with the standard faience, both the Egyptian blue frit and vitreous faience appear to have been produced in the Aegean using locally available sources of quartz, lime, alkali and colorant, and it was only glass that was imported into the Aegean as a "raw material", most probably from Egypt in the case of cobalt blue glass.

In trying to explain this difference, the first point to emphasise is that faience objects and, if glazed, vitreous faience objects must be produced direct from quartz, alkali and colorant and an intermediate "raw material" stage is not an option. In contrast, it is generally assumed that, as with glass, Egyptian blue objects were produced in two stages via an intermediate "raw Egyptian blue frit" stage (Tite et al 1984). Therefore, the fact that glass alone was imported as a "raw material" was perhaps due to the fact that glass production was more centrally controlled. This central control could have been because glass production was a more complex technology than that of Egyptian blue. Alternatively, it could have been because glass, like lapis lazuli and turquoise, was seen as a semiprecious stone that came from far away and exotic lands, and that had supposedly magical properties.

Most of the faience seems to have been made in Crete or by Minoan craftpersons but from the end of the 15th century the balance of production, first of faience, and then mainly of Egyptian blue frit, glass and vitreous faience is shifted to the mainland. It may sound oversimplified to say that faience-making was a Minoan centred and directed craft while glass-making (especially relief beads and plaques) was Mycenaean centred and directed as may also have been Egyptian blue frit and vitreous faience-making and the use of cobalt blue colorant. This idea is supported by the sheer quantities of faience objects found at the Palace of Knossos and the quantities of Egyptian blue beads, glass beads and plaques and vitreous faience beads from the Mycenaean tombs. Further support for this proposed shift in the principal production centre is perhaps provided by the apparent corresponding change in the type of alkali used. Thus, on the very limited evidence available, a potash rich plant ash appears to have been used in the production of Minoan faience and the earliest Egyptian blue frit from the Vat Room Deposit at Knossos. In contrast, a soda rich plant ash or perhaps an evaporite such as natron was used for the post 1500 BC vitreous faience produced when the mainland was the dominant centre for the production of vitreous materials.

To extend our understanding of the production technology of Aegean vitreous materials, the following further work is in progress:

- finalise quantitative analyses for the present group of samples and extend the range of vitreous materials analysed to include samples from the Mycenaean mainland;

– explore possibility of obtaining semi-quantitative non-destructive analyses using LIBS and XRF, and thus extending significantly the range of samples available for analysis;
– complete faience replications and extend to include the replication of vitreous faience and glass;
– obtain compositional data for plant ashes from Crete and the mainland, and investigate the possibility that there was an Aegean (or Macedonian) source of natron;
– locate possible sources of cobalt in the Aegean, and undertake trace element analysis in an attempt to identify the source used in the production of vitreous faience.

REFERENCES

Alexiou, S., 1961–2, Οι Πρωτομινωικοί τάφοι της Λεβήνος και η εξέλιξις των προανακτορικών ρυθμών, *Kritika. Chronika* 1: 88–91.

Alexiou, S., 1967, *Υστερομινωικοί τάφοι λιμένος Κνωσού (Κατσαμπά)*. Arhaeologiki Etairia. Athens.

Andreopoulou-Mangou, E., 1988, Chemical Analysis of Faience Objects in the National Archaeological Museum, *New Aspects of Archaeological Science in Greece BSA Occasional Paper 3 of the Fitch* Laboratory: 15–18.

Andrews, C., 1990, *Ancient Egyptian Jewellery*. The British Museum. London.

Anglos, D., 2001, Laser-induced breakdown spectroscopy in art and archaeology, *Applied spectroscopy*, 55.6, 186–205.

Bass, G.F., 1991, Evidence of Trade from the Bronze Age Shipwrecks, N.H. Gale (ed) *Bronze Age Trade in the Mediterranean*. Papers presented at the Conference Held at Rewley House, Oxford in December 1989, *SIMA* 90: 69–82.

Beck, C.H., 1928, Classification and Nomenclature of Beads and Pendants, *Archaeologia* 77, 1–76.

Bellintani, P., 2003, Quali quante conterie: perle ed altri materiali vetrosi dell' Italia settentrionale nel quadro dell' eta del Bronzo Europea, in *Atti Della XXXV Riunione Scientifica, Le comunità della Preistoria Italiana studi richerche sul Neolitico e le Età Dei Metalli, IInstituto Italiano Di Preistoria, In memoria di Luigi Bernabò Brea*, 483–98, Firenze.

Blegen, C.W. and Pierce-Blegen, E., 1937, *Prosymna: The Helladic Settlement Preceding the Argive Heraeum*. Cambridge University Press, Cambridge.

Branigan, K., 1974, *Aegean Metalwork of the Early and Middle Bronze Age*, Clarendon Press. Oxford.

Brill, R.H., 1999, *Chemical analyses of early glass*, Corning Museum of Glass, New York.

Cadogan, G., 1976, Some faience, blue frit and glass from fifteenth century Knossos, *Temple University Aegean Symposium* 1: 18–19.

Cadogan, G., 1981, A Probable Shrine in the Country House at Pyrgos, R. Hägg and N. Marinatos (eds) *Sanctuaries and Cults in the Aegean Bronze Age*. Proceedings of the First International Symposium at the Swedish Institute in Athens, 11–13 May 1980: 169–172.

Cadogan, G., 2001, Catalogue Entries, A. Karetsou, M. Andradaki-Vlazaki, N. Papadakis (eds.), *Crete-Egypt. Three thousand years of cultural links*, Ministry of Culture. Herakleion-Cairo.

Cameron, M.A.S., 1987, The Palatial Thematic System in the Knossos Murals in R. Hägg and N. Marinatos (eds) *The Function of the Minoan Palaces*. Proceedings of the Fourth

International Symposium at the Swedish Institute in Athens, 10–16 June, 1984, 321–328.
Chatzi-Spiliopoulou, G., 2002, Μυκηναϊκό γυαλί, in G. Kordas and A. Andonaras (eds) Ιστορία και τεχνολογία του Αρχαίου γυαλιού, Glasnet, Athens: 63–87.
Cline, E.H., 1994, *Sailing the Wine-Dark Sea. International Trade and the Late Bronze Age Aegean. BAR International Series* 591. Oxford.
Crowell, B., 1998, Catalogue Entries in F. D. Friedman, (ed) *Gifts of the Nile. Ancient Egyptian Faience*. Thames and Hudson. London.
Demakopoulou, K., 1998, *Ο Θησαυρός των Αηδονίων. Σφραγίδες και κοσμήματα της Ύστερης Εποχής του Χαλκού*, Ministry of Culture. Athens.
Detournay, B., Poursat, J.-Cl., Vandenabeele, F.,1980, *Fouilles exécutées à Mallia. Le Quartier Mu II, ÉtCrét* 26. Paris.
Doumas, C., 1992, *The Wall-Paintings of Thera*, Idrima Theras P. Nomikos. Athens.
Evans, A.J., 1906, The Prehistoric Tombs of Knossos, *Archaeologia* 59, 391–562.
Evans, A.J., 1921–1935, *The Palace of Minos at Knossos* I-IV, Macmillan. London.
Foster, K.P., 1979, *Aegean Faience of the Bronze Age*, Yale University Press. New Haven & London.
Foster, K.P., Kaczmarczyk, A., 1982, X-Ray Fluorescence Analysis of some Minoan faience, *Archaeometry* 24, 143–157.
Foster, K.P., 1987a, Composition of colours in Minoan faience, in M. Bimson and I.C. Freestone (eds), *Early Vitreous Materials*. British Museum Occasional Papers 56, 57–64.
Foster, K.P., 1987b, Reconstructing Minoan palatial faience workshops, in R. Hägg and N. Marinatos (eds), *The Function of the Minoan Palaces*. Proceedings of the Fourth International Symposium at the Swedish Institute in Athens, 10–16 June, 1984, 11–16.
Frankfort, H., (1989) 1996, *The Art and Architecture of the Ancient Orient*, Yale University Press. New Haven & London.
Friedman, F.D., (ed) 1998, *The Gifts of the Nile. Ancient Egyptian Faience*, Thames and Hudson in association with the RISD Museum of Art, Rhode Island School of Design, Singapore.
Frödin, O. and Persson, A., 1938, *Asine: Results of the Swedish Excavations 1922–1930*, Generalstabens Litografiska Anstalts Förlag. Stockholm.
Grose, D., 1989, *Early Ancient Glass: core-formed, rod-formed, and cast vessels and objects from the Late Bronze Age to the Early Roman Empire. 1600 BC – 50 AD*, Hudson Hills Press in association with the Toledo Museum of Art. New York.
Haevernick, T.E., 1960, Beiträge zur Geschichte des antiken Glases, III, Mykenisches Glas, *Jahrbuch des Römisch–Germanischen Zentralmuseums Mainz* 7, 36–53. Reproduced in Haevernick 1981, 71-83.
Haevernick, T.E., 1963, Mycenaean glass, *Archaeology* 16.3, 190–93.
Haevernick, T.E. 1981, *Beiträge zur Glasforschung: die wichtigsten Aufsätze von 1938 bis 1981*. Verlag von Zabern. Mainz.
Hayes, W.C., 1990, *The Scepter of Egypt*. The Metropolitan Museum of Art, New York.
Hood, S.M.F., 1978, *The Arts in Prehistoric Greece*, Pelican, Harmondsworth.
Hughes-Brock, H., 1998, Greek Beads of the Mycenaean Period, in L.D. Sciama and J.B. Eicher, (eds) *Beads and Bead Makers: Gender, Material Culture and Meaning*, Oxford and New York: 247–71.
Hughes-Brock, H., 1999, Mycenaean Beads: Gender and Social Context, *OJA* 18.3, 277–296.
Ignatiadou, D., Dotsika E., Kouras A., Maniatis Y., 2004 (in press), Nitrum Chalestricum.

The natron of Macedonia. Paper presented at the 16th Congrès de l'Association Internationale pour l'Histoire du Verre. London 7th–13th September, 2003.

Karo, G., 1930, *Schachtgräber von Mykenai*, Bruckmann. München.

Kaczmarczyk, A. and Hedges, R.E.M., 1983, *Ancient Egyptian Faience: An Analytical Survey of Egyptian Faience from Predynastic to Roman Times*, Aris & Phillips. Warminster.

Kaczmarczyk, A., 1986, The source of cobalt in ancient Egyptian pigments, in J.S. Olin and M.J. Blackman (eds), *Proceedings of the 24th International Archaeometry Symposium*, Smithsonian Institution Press, Washington, DC, 369–76.

Lacovara, P., 1998, Nubian Faience, in F.D. Friedman (ed), *Gifts of the Nile. Ancient Egyptian Faience*, Thames & Hudson in association with the RISD Museum of Art, Rhode Island School of Design. Singapore, 46–49.

Lembese, A., 1967, Ανασκαφή τάφου εις Πόρον Ηρακλείου, *Praktika tis Archaeologikis Etairias*, 195–209.

Leveque, M., 1998, Catalogue Entries, in F.D. Friedman (ed), *The Gifts of the Nile. Ancient Egyptian Faience*. Thames and Hudson in association with the RISD Museum of Art, Rhode Island School of Design. Singapore.

Levi, D., 1961–1962, La tomba a tholos di Kamilari presso a Festòs, *ASAtene* 39–40, 7–148.

Lilyquist, C., 1993, Granulation and Glass: Chronological and Stylistic Investigations at Selected Sites, ca. 2500-1400 B.C., *BASOR* 290–291, 29–94.

Lilyquist, C. and Brill, R.H., 1995, *Studies in Early Egyptian Glass*, The Metropolitan Museum of Art. New York.

Lolling, H.G., 1880 *Das Kuppelgrab bei Menidi*, Deutsches Archäologisches Institut in Athen. Athens.

Lucas, A. and Harris, J.R., 1962, *Ancient Egyptian Materials and Industries*, Histories and Mysteries of Man. London.

Mirtsou, E., Vavelidis, M., Ignatiadou, D., Pappa, M., 2001, Early Bronze Age faience from Agios Mamas, Chalkidiki: a short note, in Y. Basiakos, E. Aloupi, Y. Facorellis (eds), *Archaeometry Issues in Greek Prehistory and Antiquity*, Hellenic Society of Archaeometry, Society of Messenian Archaeological Studies. Athens, 309–16.

Money-Coutts, M.B., 1935-1936, The Cave of Trapeza: Miscellanea. *BSA* 36, 122–25.

Muhly, P., 1992, Μινωικός λαξευτός τάφος στον Πόρο Ηρακλείου. Arhaeologiki Etairia. Athens.

Müller, K., 1909, Alt – Pylos II: Die Funde aus den Kuppelgräbern von Kakovatos, *AthMitt* 34, 269–328.

Oates, D., Oates, J., McDonald, H., 1997, *Excavations at Tell Brak: vol. 1: The Mitanni and Old Babylonian periods*. BSAI.

Nightingale, G., 2000, Mycenaean Glass Beads: jewellery and design in *Annales du 14e Congrès de l' Association Internationale pour l' Histoire du Verre*, Venezia – Milano 1998. Lochem, 6–10.

Nightingale, G., 2002, Aegean Glass and Faience Beads: An Attempted Reconstruction of a Palatial Mycenaean high-tech Industry, in G. Kordas (ed), *Hyalos, Vitrum, Glass. History, Technology and Conservation of Glass and Vitreous Materials in the Hellenic World. 1st International Conference*, Glasnet. Athens, 47–54.

Panagiotaki, M., 1995, Preliminary technical observations on Knossian faience, *OJA* 14.2, 137–149.

Panagiotaki, M., 1999a, Minoan Faience-and Glass-making: Techniques and Origins, in P.P. Betancourt, V. Karageorghis, R. Laffineur and W-D. Niemeier (eds), *Meletemata*.

Studies in Aegean archaeology presented to Malcolm H. Wiener as he enters his 65th year. *Aegaeum* 20, 617–622.

Panagiotaki, M., 1999b, *The Central Palace Sanctuary at Knossos*, BSA suppl. Vol. 31. Great Britain.

Panagiotaki, M., 2000, Crete and Egypt: contacts and relationships seen through vitreous materials in A. Karetsou (ed) *Κρήτη-Αίγυπτος. Πολιτισμικοί δεσμοί τριών χιλιετιών*. Vol. 1 (Meletes), Ministry of Culture. Athens, 154–61.

Panagiotaki, M., 2001, Catalogue Entries, in A. Karetsou, M. Andreadaki-Vlazaki, N. Papadakis, (eds), *Crete-Egypt*. Three thousand years of cultural links (Catalogue), Ministry of Culture. Herakleion-Cairo.

Panagiotaki, M., 2002, Φαγεντιανή – Κύανος – Ύαλος: Ύλες των βασιλέων, των Θεών και των νεκρών της αρχαιότητας, in G. Kordas and A. Andonaras (eds), Ιστορία και τεχνολογί α του Αρχαίου γυαλιού. Glasnet. Athens, 33–62.

Panagiotaki, M., Sklavenitis, C., Maniatis, Y., Tite, M.S. (in press), Experiments in making Vitreous Materials. Paper presented at the 9th International Congress of Cretan Studies at Elounda from 1st to 6th October 2001.

Panagiotaki, M., Papazoglou-Manioudaki, L., Chatzi-Spiliopoulou, G., Maniatis, Y., Tite, M.S., Shortland, A., 2004 (in press), A glass workshop at the Mycenaean citadel at Tiryns in Greece. Paper presented at the 16th Congrès de l'Association Internationale pour l'Histoire du Verre, London 7th-13th September 2003.

Panagiotaki, M., et al. (forthcoming), *Aegean Vitreous Materials: Manufacturing techniques*.

Peltenburg, E.J., 1972, On the Classification of Faience Vases from Late Bronze Age Cyprus, *Praktika tou Protou Diethnous Kiprologikou Synedriou*, 129–136.

Peltenburg, E.J., 1974, Appendix I: The Glazed Vases (including a polychrome rhyton) with analyses and technical notes by H. Mckerrell, V. Karageorghis (ed) *Excavations at Kition I*, 105-144, Department of Antiquities of Cyprus. Nicosia.

Peltenburg, E.J., 1991, Greeting Gifts and Luxury Faience: A context of Orientalising Trends in Late Mycenaean Greece, in N.H. Gale (ed) *Bronze Age Trade in the Mediterranean SIMA 90*, 162–179.

Phillips, J.S., 1991, *The Impact and Implications of the Egyptian and 'Egyptianizing' Material Found in Bronze Age Crete ca. 3000–ca. 1100 BC*. (PhD diss.) University of Toronto.

Platon, N., 1974 *Ζακρος Το νεον Μινωικον ανακτρον*, Archaiologiki Etairia. Athens.

Platon, L., 1993, Ateliers palatiaux minoens: une nouvelle image, *BCH* 117, 103–122.

Persson, A.W., 1931, *The Royal Tombs at Dendra near Midea*, Gleerup. Lund.

Sakellarakis, J., 1990, The fashioning of Ostrich-egg Rhyta in the Creto-Mycenaean Aegean, in D.A. Hardy (ed) *Thera and the Aegean World III*, 285–308. The Thera Foundation. London.

Savignoni, L., 1904, Scavi e scoperte nella necropolis di Phaistos, *MonAnt*, 14, 501–666.

Sayre, E.V and Smith, R.W, 1974, Analytical studies of ancient Egyptian glass, in A. Bishay, (ed), *Recent advances in the science and technology of materials: Volume 3*, Plenum Press. New York, 47–70.

Seager, R.B., 1912, *Exploration in the Island of Mochlos*, The American School of Classical Studies. Boston-New York.

Shortland, A.J., 2000, *Vitreous materials at Amarna*, BAR International Series 287. Oxford.

Spaer, M., 2001, *Ancient Glass in the Israel Museum: Beads and Other Small Objects*, The Israel Museum. Jerusalem.

Stern, E.M. and Schlick-Nolte, B., 1994, *Early Glass of the Ancient World, 1600 BC–AD 50*, Ernesto Wolf Collection. Ostfildern.

Tite, M.S., Bimson, M. and Cowell, M.R., 1984, Technological examination of Egyptian blue, in J.B. Lambert (ed), *Archaeological Chemistry III*, American Chemical Society Advances in Chemistry Series No.205, Washington, D C, 215–242.

Tite, M S and Bimson, M, 1986, Faience: an investigation of the microstructures associated with the different methods of glazing, *Archaeometry* 28, 69–78.

Tite, M S and Shortland, A J, 2003, Production technology for copper and cobalt blue vitreous materials from the New Kingdom site of Amarna – a reappraisal, *Archaeometry*, 45, 273–300.

Tournavitou, I., 1995, *The 'Ivory Houses' at Mycenae*, BSA Suppl. 24. London.

Tournavitou, I., 1997, Moulds and Jewellers Workshops in Mycenaean Greece. An Archaeological Utopia, in C. Gillis, Chr. Risberg and B. Sjoberg (eds), *Trade and Production in Premonetary Greece. Production and Craftsman*. Proceedings of the 4th and 5th International Workshops. Athens 1994 and 1995, SIMA Pocket book 143, 211–230.

Triantafyllidis, P., 2000, Ροδιακή Υαλουργία I. Τα εν θερμώ διαμορφωμένα διαφανή αγγεία πολυτελείας. Ministry of the Aegean. Athens.

Vandiver, P., 1982a, Technological Change in Egyptian Faience, in J.S. Olin and A.D. Franklin (eds), *Archaeological Ceramics*. Smithsonian Institution Press, Washington DC, 167–179.

Vandiver, P., 1982b, Mid-Second Millennium B.C. Soda-Lime-Silicate Technology at Nuzi (Iraq), in T. A. Wertime and S. F. Wertime (eds), *Early Pyrotechnology*, Smithsonian Institution Press, Washington DC, 73–92.

Vandiver, P., 1983, The Manufacture of Faience, App. A., in A. Kaczmarczyk and R.E.M. Hedges: *Ancient Egyptian faience: An analytical survey of Egyptian faience from Predynastic to Roman times*, Aris & Phillips. Warminster, 1–144.

Vandiver, P B, 1998, A review and proposal of new criteria for production technologies of Egyptian faience, in S. Colinart, & M. Menu (eds), *La couleur dans le peinture et l'émaillage de l'Égypte ancienne*, Edipuglia. Bari, 121–139.

Xanthoudides, St., 1924, *The Vaulted Tombs of the Messara. An account of some early cemeteries of southern Crete*, University Press of Liverpool. London.

Xenaki-Sakellariou, A., 1985, Οι θαλαμωτοί τάφοι των Μυκηνών ανασκαφής Χ. *Τσούντα (1887-1898)*, Diffusion de Boccard. Paris.

Wace, A., 1956, Mycenae 1939–1953: Part I: Preliminary Report of the Excavations of 1955, *BSA* 51, 103–122.

Warren and Hankey, V., 1989, *Aegean Bronze Age Chronology*, Bristol Classical Press. Bristol.

Weinberg, G.D., 1961–62, Two glass vessels in the Heraklion Museum, *Kritika Chronika* 15, 226–29.

Chapter 9

Egyptian Sculptors' Models: functions and fashions in the 18th Dynasty

Sally-Ann Ashton

Abstract

'Sculptors' models, 'artists' trial pieces' and 'ex-votos' are all terms which are used to describe a disparate group of unfinished reliefs and partial sculptures in-the-round. This paper will consider the frequency, use and types of material from Eighteenth Dynasty Egypt in an attempt to establish whether such objects were introduced because of stylistic, religious or social developments during the reign of Akhenaten, or, whether they are part of a wider phenomenon. The material will be divided into categories and considered in terms of find-spots and possible functions. An initial survey of the sub-groups suggests that models were produced at specific times, when there had been substantial changes in iconography and style of the royal image, but also later in the 4th and early 3rd centuries B.C. Both periods saw considerable developments in religion and more specifically in the worship of the members of the ruling house, which may explain why the phenomenon is limited. This paper will consider early works on models from the Amarna period, most notably Aldred (1973) and Arnold (1996), along with unpublished models from the Petrie Museum of Egyptian Archaeology, London and the Fitzwilliam Museum, Cambridge in order to reassess the function of this important group of objects. The Ptolemaic examples from both museums will form a forthcoming monograph.

Early in the reign of Amenhotep IV an artistic revolution occurred in the representation of the king of Egypt. Colossal statues erected at the temple of Aten in Karnak showed the ruler and his wife not in the usual idealised manner, but with hugely exaggerated features, which to many early scholars seemed to defy the ideology of kingship (Figure 9.1). However, this was not the first time in Egyptian history that a non-idealised, and often erroneously called 'realistic' type of image was chosen to represent the king. In the 12th and 13th Dynasties the rulers were shown with portraits that adhered to the so-called 'veristic' style, with wrinkles and lines depicting age on the face. The early representations of

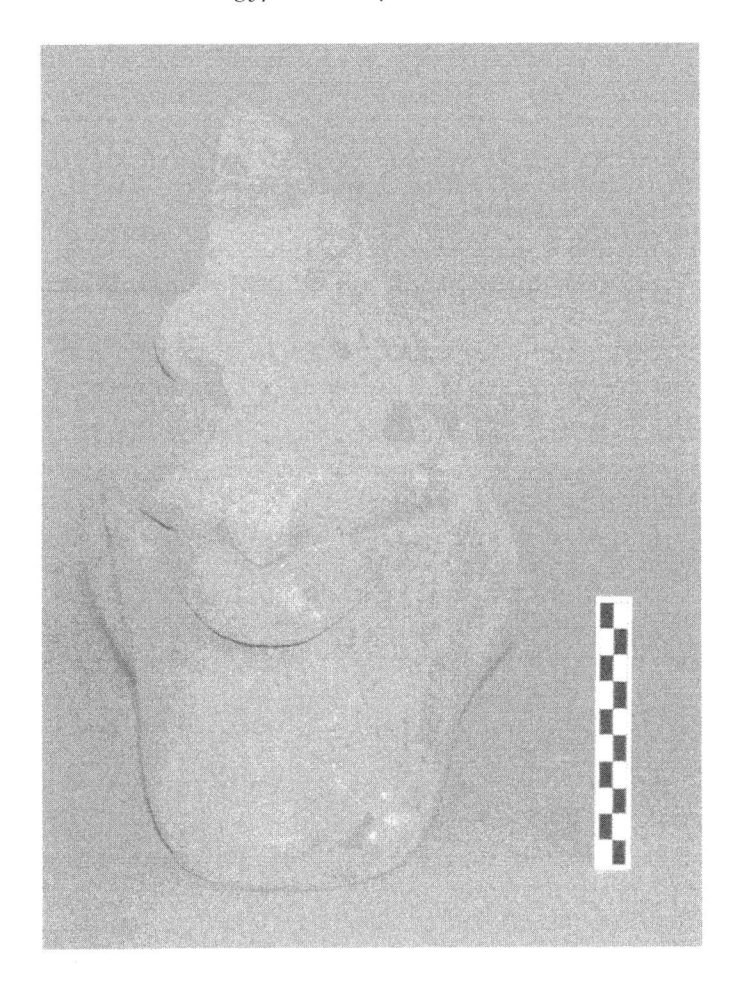

Figure 9.1 EGA.4516.1943 (Copyright and reproduction courtesy of The Fitzwilliam Museum, Cambridge).

Amenhotep IV/Akhenaten, however, took the royal image one step further away from the idealised in that the bodies were also 'distorted' with regard to the canons of Egyptian art, revealing a distended stomach, voluptuous hips and rounded thighs. Fingers and toes also became elongated, creating new proportions for the extremities that mimicked the portraits and bodies of the rulers. Such changes in the representation of the pharaoh occurred at times of social, political or religious change, often when there is a need to re-define either the role or status of the dynasty or individual ruler. In the cases of the royal women, socio-political developments were often designated by the adoption of a new or specific attribute, as the 'portrait' features typically mimicked those of the king or remained idealised in form.

In a world where artistic expression was denied, a world that seems, to the modern western viewer, rigid in terms of its three thousand-year history of artistic representations, such innovations appear at first to be incongruous. However, throughout Egyptian history the artists show that they are easily able to adapt to new styles and indeed even outside influences, which were adopted during periods of foreign rule. Perhaps for this reason, there are many shared features between the Amarna and Ptolemaic periods, not least the re-carving and re-use of statues from the former by artists in the latter (Lauffrey 1971, 71, Fig. 13;1979, 88–89; Bianchi 1980, 11); there are also iconographic links between representations of Amarna and Ptolemaic royal women. Both periods were also times of innovation, in terms of society and religion. The question that will be addressed in this paper is whether such artistic developments warranted changes in the way that sculptors worked, and more specifically, if these changes required models for guidance.

This hypothesis at first glance, appears to be acceptable and indeed logical, but in fact Egyptian sculptors' models from all periods are extremely problematic. There are several reasons for this, and I do not claim to have solved all of the complexities or problems but, by dividing the material into groups according to their chronological sequence and type, it is then possible to consider the wider social or religious implications behind their apparent prevalence during certain times.

The so-called models fall into two groups of material: limestone and plaster, and cover three distinct themes: royal, divine, and private. Their forms may suggest distinct functions within the main group, for there are both relief plaques, some of which are carved on both sides, and busts, which are in the round. There are also plaster 'masks,' often hollow at the back, which represent both private and royal subjects in the Amarna period, but are expanded in the Ptolemaic repertoire to include traditional Egyptian deities.

The earliest relief plaques seem to date to the reign of Amenhotep III. One such example, now in the Egyptian Museum in Berlin (21.299), shows Amenhotep III interestingly with a second uraeus at the side of the head in addition to the usual placement on the brow. This piece was found at Tell el-Amarna and was perhaps used as a model; the double uraei are unusual for a male figure, although they do appear on images of Tiye and on early representations of Nefertiti (Ashton forthcoming. The appearance of models during the reign of Amenhotep III corresponds to the roots of the developments of artistic styles under his successor.

There are two further examples from the 18th Dynasty in the Fitzwilliam Museum that I would like to mention one before looking at the Amarna period, and I use them here to illustrate an important point about student pieces or trial pieces. One is a chariot with an ear study on one side (Figure 9.2), which seems more likely to be a trial piece or re-used block because of the two unconnected subjects (one would perhaps expect a head with a detail of the ear). On the reverse are several small studies of ears and hands, all of which are roughly carved, and it appears that these, along with the ear, are secondary additions to what was once a representation of a chariot. The elongated White Crown is

Figure 9.2 EGA.3129.1943 (Copyright and reproduction courtesy of The Fitzwilliam Museum, Cambridge).

stylistically similar to that worn by Akhenaten and is paralleled on several relief representations. There is also a small head of a Nubian woman from Giza, measuring 5 cm in height and which has been re-carved at the back of the head (Fitzwilliam EGA.4550.1943, Figures 9.3–4), illustrating that stone was re-used, no matter how small a piece. Both examples have clearly been re-used probably as practice pieces, but what is noteworthy is that neither conforms to the standard plaques with royal representations.

As a general rule it seems that the relief plaques appear when there is a specific purpose in mind. There are a number that have been dated to the 25th and 26th Dynasties. There are several examples of archaising plaques, which imitate a style more commonly associated with the Old Kingdom. One such example in

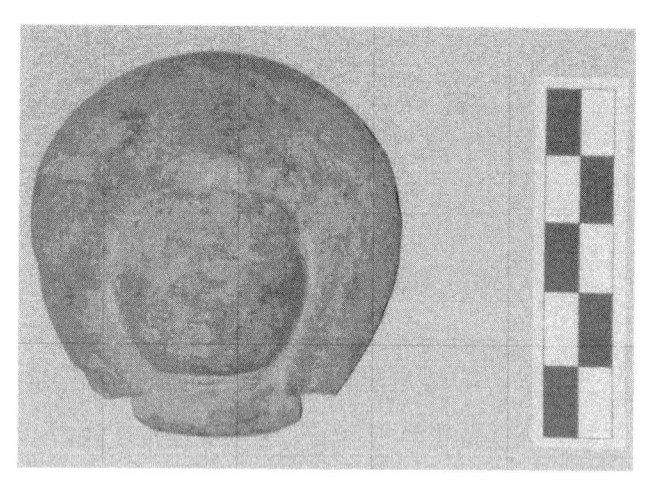

Figure 9.3–4 EGA.4550.1943 *(Copyright and reproduction courtesy of The Fitzwilliam Museum, Cambridge).*

the Fitzwilliam Museum shows the head of a man, which is archaising but which has an ear study on the reverse (Figure 9.5). There are traces of red paint on the surface, and the rear is badly scratched but is otherwise untouched. Bianchi also published other examples from this period in 1979, arguing that they were ex-votos. One particular example now in the Egyptian Museum in Cairo (CG 33419), as illustrated in Edgar (1906, 60 Pl. XXVI) is stylistically similar to the portrait of Amenhotep III on the west wall of the west chapel of Khonsu at Luxor (Berman 1990, Pl. 7c). Two profiles are preserved from what was originally four, each evenly spaced and cut into the stone in the prescribed space. The nose of the

Figure 9.5 EGA.72.1949 (Copyright and reproduction courtesy of The Fitzwilliam Museum, Cambridge).

lower example is, however, uncharacteristic of this particular ruler's relief representations, and the possibility that this is one of the many 'archaising' plaques from the Late or Ptolemaic periods cannot be dismissed, particularly when there are parallels for this form on the later models. Myśliwiec (1972, 73) dated the Cairo model to the Ptolemaic period, considering it to be a student exercise. This technique and subject can be found on another Ptolemaic example, now in the Oriental Institute Museum, Chicago (10822) and reveals an interest in art from earlier periods of Egyptian history. A similar example, although in this instance showing the repeated practice of carving hands can be found on an example in the Petrie Museum, UC 2234 (Figure 9.6). Only two original sides are preserved, the top and side to the viewer's right. There are also possible traces of red pigment on the upper edge, which may, or may not be part of the original design. Although only three hands are preserved (all left hands), it seems possible that there was a fourth. The style is not typical of the Amarna period, and fits well within the Ptolemaic repertoire.

By far the largest group and the closest in terms of evidence of types for the Amarna material, come from the Ptolemaic period. Royal busts, plaque reliefs and plaster heads have been found at sites such as Memphis and it has been argued that they were used in order to disseminate a royal image that was visually similar to that of the 30th Dynasty rulers (Ashton 2001, 19–20, 82–3, No.3). This was a political policy that was replicated in the choice of temples at which the Ptolemies made dedications and is shown nowhere more clearly than in the portraits in the round, which are virtually indistinguishable from their 30th Dynasty counterparts.

182   *Sally-Ann Ashton*

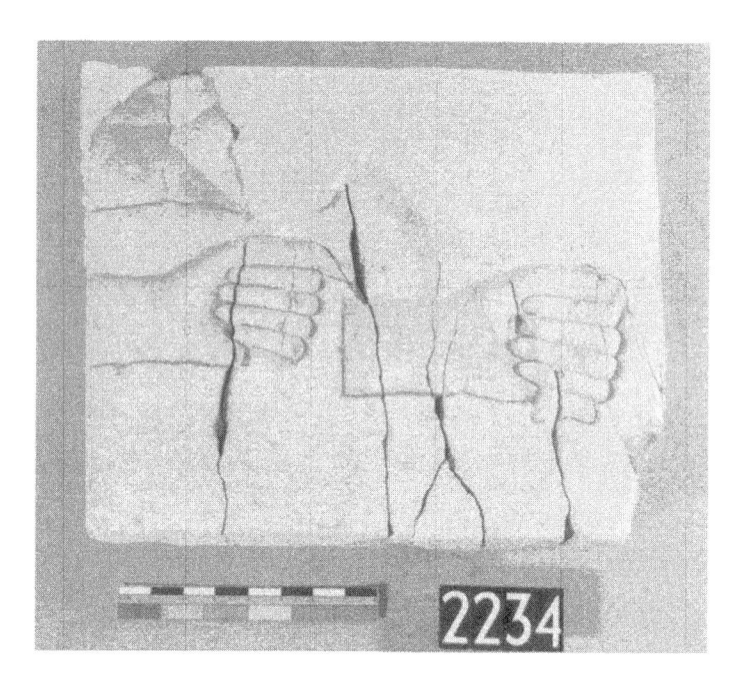

Figure 9.6 UC 2234 (Copyright and reproduction courtesy of The Petrie Museum of Egyptian Archaeology, London).

Considerably more work has been undertaken on sculptors' models from the 4th and 3rd centuries B.C. Here, the debate has concentrated on two points, firstly whether the models functioned solely in this capacity, or if they were in fact ex-votos. Scholars writing on the subject are divided. Secondly, and to a lesser degree, the date of these images has been questioned. The question of whether they are 30th Dynasty or Ptolemaic has been raised and not satisfactorily decided by stylistic comparison with sculptures of the two periods, where there is a similar problem over identifying rulers due to fragmentary evidence and a general lack of inscriptions (Josephson 1997). The second of these issues is not relevant for the present paper, but the question of function and also reasons for the appearance of the so-called models is pertinent to the discussion of the Amarna material.

Although more work has been undertaken on the Ptolemaic examples, the majority are without a definite provenance and so it is extremely difficult to know how they were intended to be used. To make matters worse, even those with a provenance such as the Petrie Museum Memphis examples, could have come from either a workshop or a temple environment (Ashton 2003, 59–64). One other example in the Louvre was found in a private house at Edfu, and it has been suggested this was the house of a sculptor (museum records). The Ptolemaic royal family, however, were worshipped as gods and so it is possible that the piece, which has a representation of a Ptolemaic queen carved onto one side, was used as part of a private shrine. Some plaques have holes drilled either right through (in the case of the Amarna material) or into the back (in the case of the Ptolemaic examples). It is not, however, clear that the drill-holes in the backs of the Ptolemaic examples are ancient (Ashton 2003, 62–63). This problem aside, the existence of a hole does not solve the additional puzzle over their function because just as a plaque could hang in a temple, so it could also hang in a studio. Smaller stelae such as the ear stelae do not have holes and these at least we can probably assume were votives. The question over who exactly used such plaques remains open. There are of course several possibilities. Presuming that the master sculptor carved these images (and Aldred does not) they could be taken to the building that was to be decorated and copied, but this raises questions: was a plan drawn onto the wall prior to carving, and did the same person draw the scheme and then carve it, or were the tasks undertaken by two individuals? There is interestingly a general lack of evidence for the use of pigment prior to carving on the plaques, but the question of the number of craftsmen involved is an interesting one, that may ultimately affect our interpretation of this material. Models in the round are less problematic and some will be discussed in the current paper in order to illustrate the possible uses of the relief plaques.

The find-spots of some of the Amarna period models were more carefully recorded than the Ptolemaic pieces from Memphis, and as a consequence this material is of considerable importance not only in its own right, but also for other periods. There are certain criteria that must be met before this group of material can be considered within the wider historical context. Firstly, it is necessary to establish when a piece is a model, and when it is simply an unfinished sculpture or a part of the manufacturing process. Timing is also an important factor; one would expect models for copying to appear during periods of change or when there was a desire to express continuity.

RELIEF PLAQUES

The first group is perhaps the most obvious candidate to support the idea that models were needed to either disseminate or practice the new royal image. They could, in theory also be connected with the role of the royal couple Akhenaten and Nefertiti in the worship of the Aten. These are plaque-like relief 'models' from Tell el Amarna. In his *Akhenaten and Nefertiti* exhibition and catalogue, Aldred discussed five trial pieces, which he dated to the early period on account of their non-idealised features (Aldred 1973, Cats. 9–10, 12, 15, 38). All of the selected examples show Akhenaten and all have the exaggerated features that are typical of this period, but Aldred distinguished between those that he considered to be student pieces (Cats. 9–10, 38), and one produced by a master sculptor (Cat. 12) and one unfinished piece (Cat. 15), which he attributes to a 'skilled and experienced pupil.' Already, by making these distinctions, Aldred defined the perimeters of their function. For, if we assume that junior sculptors carved these

Museum	Number	H. W. Th. (cm)	Period	Re-used?	Main subject	Aldred reference.
Brooklyn, BMA	67.175.1	16.7, 18, 3.7	early	yes	Akhenaten and Nefertiti	38
Brussels, MRAH	E. 3052	15.2, 13.1, 1.7	early	yes	Akhenaten	10
Brussels, MRAH	E. 3051	13.9, 9.7, 2.6	early	no	Akhenaten	9
London, Petrie	UC 087	19, 15, 2-4.6	early	No	Akhenaten	-
New York, MMA	69.99.40	34.8, 23.4, 4	early	no	Akhenaten	15
Cairo, Egyptian	JE 59296	27, 16.5, 4	early	possibly	Nefertiti/kneeling man	-
Edinburgh, Royal Scottish	1969.377	26.3, 21.3, 4.3	mid?	yes	Akhenaten	12
London, Petrie	UC 011	8.7, 7.5, 1.2	mid	no	Nefertiti?	59
London, Petrie	UC 013	20.5, 14.5, 3.2	mid-late	yes	Akhenaten	-
Cairo, Egyptian	JE 59294		mid	possibly	Akhenaten	Fig. 49
Brooklyn, BMA	16.48	15, 22.1, 4.2	Late	no	Akhenaten and Nefertiti	121
New York, MMA	21.9.13	17.8, 14, 3.1	Late	yes	Akhenaten	119
Berlin, ÄM	DDR 21683	21.4, 21, 4.1	Late	no	Akhenaten	115
London, Petrie	UC 036	10.2, 6.5, 3.3	Late	yes	Akhenaten	-
London, Petrie	UC 402	18.4, 18.7, 4	Late	no	Nefertiti	-

Table 9.1 showing a summary of Amarna relief plaques.

pieces, then they fall into the category of trial piece rather than model, and yet a very similar, but in Aldred's opinion, better standard image of Akhenaten wearing the Blue Crown (Cat. 12), is categorised as a model.

In fact, this example is no more an artistic masterpiece than the others, it is simply complete. In my opinion, there is little evidence to support the division between master and pupil; the reliefs are roughly the same size and proportions, and they all show the same subject matter, the king. The only difference is that there is negative space on the unfinished piece, suggesting that the form of a model was still to be established. This observation may not ultimately change Aldred's hypothesis of their use, but it is important to establish whether these plaques were trial pieces or whether they served as models for other artists copying the official royal style. The attention to specific details on the main image might in fact suggest that they were working models. In other words, that specific features were highlighted by the master sculptor, or, with nothing else around the artist used the surface or in some instances re-used the back to practice. I would like to make a distinction between master and student, and suggest that the latter used cast-offs as illustrated by the re-used limestone block decorated with a chariot and ear on one side, and a variety of crudely incised figures on the rear (Figure 9.2). This distinction will be supported by the re-use of what appear to be formal models, with less skilfully executed additions to the main image. I will also suggest that some of the relief representations below acted as trial pieces for the models used by master sculptors, thus adding to Aldred's original categories.

The first relief plaque (Aldred 1973, Cat. 38) is said to have come from a house in Tell el-Amarna, and shows two heads facing each other; that on the viewer's left shows the shoulders whereas the figure on the right is pushed up against a later break. Above is the profile of a hand. Aldred described the piece as a 'student exercise' because of the poor quality of the carving. The double portraits are found on two other examples that I will address shortly, and some scholars have applied a very different interpretation to this specific form of plaque. The fact that the sides of the plaque are damaged and that these breaks appear to be ancient (Aldred 1973, 120) might suggest that the object which originally functioned as a model, may have been re-used in antiquity. It seems likely that the two heads were part of the original scheme, because if one had been intended it would have been placed in a vertical position on the plaque. The hand, however, could be a sketch by the artist using the model (for the elongated fingers certainly place it within the Amarna period), or, it may have been re-used at another time. The significance of the two figures facing each other might suggest that this piece functioned as a model for a composite group. Finally the thickness, 3.7 cm, is typical of the relief plaque models.

Similarly, Aldred's Cat. 10 may also represent a re-used model, for it is carved on both sides. Here, the original block had broken (evident through the partial preservation of the profile), and so there is no doubt that the block was re-used. Aldred suggested that this is a trial piece, executed by a student; however, the smaller complete, and so secondary, head is actually as well carved as the original. The plaque is also thinner than most, measuring 1.7 cm as opposed to the more typical 3–4.5 cm; but it is possible that part of the original surface was re-used and then prepared for the second image. The quality and nature of carving, however, do not preclude this piece having functioned as an artist's model in both instances. This particular observation is pertinent to another question regarding the so-called models. Re-carving or re-use suggests that these images were not deemed sacred. With regard to the ex-voto versus model/trial piece debate it is not possible, as many would like to suggest, that such representations functioned as both, and whilst it is often thought that a divine image would have to be displayed in a temple, this was clearly not necessarily the case.

By contrast Aldred's Cat. 9 is an irregularly shaped plaque, measuring 2.6 cm in thickness, which is still thinner than most of the examples. A single head occupies the full space of the surface, as one would expect on a model. The wig is unfinished, and this along with the irregular shape led Aldred (1973, 96) to conclude that this was a trial piece. The uraeus and front section of the neck are scratched onto the surface rather than sketched on with ink, as found on a plaque preserved in the Petrie Museum (Aldred 1973, 136 Cat. 59). There is also some negative space below the neck and this, along with Aldred's observations and the fact that unfinished features are scratched rather than drawn onto the surface would support the hypothesis that this is a trial or student piece.

This observation is also true of another early sketch on a limestone plaque in the Petrie Museum, UC087 (Figure 9.7). Although this is a relatively large block, none of the original edges survive. The image represents a partially carved profile of a head with exaggerated features, thus placing the piece within the early period of Amarna sculpture. There are traces of red pigment on the surface and

Figure 9.7 UC 087 (Copyright and reproduction courtesy of The Petrie Museum of Egyptian Archaeology, London).

black on the bottom and viewer's right-hand edge. There are also traces of further pigment on the unevenly finished back of the plaque. The uneven depth and survival of pigment (if we assume that these are not secondary) indicate that the block was not evenly finished before the sculptor carved the image. The sketchy appearance and floating position of the head suggest that this was not an official model, or that it predates the use of such plaques. The standard of carving is not comparable to some of the later plaques, which again might suggest that the artist was not the master sculptor. This observation is particularly true of the lower neck and chin, which has been carved more deeply but not as carefully as many of the formal plaques. An assessment of this piece suggests that it is therefore a sketch or trial piece and not an official plaque. These conclusions are not however reached on account of the standard of execution alone.

A similar head, which floats in the middle of a more regularly shaped plaque, with the more standard 4 cm thickness that other models share, may be an early model (Aldred 1973, 101 Cat. 15). It is believed to have been found in a sculptor's workshop north of the Great Palace, but arrived at the Metropolitan Museum, New York via the Amherst and Gallatin collections rather than straight from Petrie's excavations. This example is categorised by Aldred as unfinished, which would suggest that it was eventually intended to fulfill a specific function. Aldred concluded that although the sculptor was in this case skilled, details such as furrows on the neck and face indicate that he was not a master craftsman. We may be seeing selective representation of a basic form. The style of this piece is without doubt early, and although the head does not fill the entire space, it is positioned centrally. This may be one of the earliest surviving models (executed before the more formal type was adopted), a status confirmed by its believed find-spot.

One early example of a relief, which has been categorised as a plaque model, came from a temple context (Cairo Egyptian Museum, JE 59296). This interesting piece was found in the foundations of the temple of the Aten at Tell el Amarna (Arnold 1996, 67, 70, 135). On one side is the portrait head of Nefertiti. In terms of its style it can be associated with the early images of the queen because of the severe style of execution, as noted by Arnold, and also because the use of the double uraeus is associated with the queen's early representations (Ashton forthcoming. On the reverse is a kneeling man with arms raised in adoration. Arnold concludes that this piece is an important link between the carvers of relief representations and the workshop of Thutmose, even though the famous bust of Nefertiti is of a later type. There is in fact a stylistically closer image of the queen on a plaque now in the Petrie Museum (UC 402, below), which supports Arnold's hypothesis. Are we to conclude that the relief plaque was formerly dedicated in the temple foundation deposit, or that it was simply left by sculptors working on the relief decoration of the temple? Of the two scenarios I would suggest the latter, simply because there is no dedication. The piece could of course have had an important association with the craftsmen who carved the temple and could have been left intentionally, but it seems fair to conclude that the principal function of a relief plaque was that of model rather than ex-voto; and the fact that it was re-used supports this conclusion. The model is important for another reason. It is one of the few formal models from the early period, in that the artist has utilised the entire space of the surface. This feature, along with its find-spot allows us a clearer idea of how such images were used, and also how they were discarded. It is however unusual in that there is a considerable negative space above the head of the queen, and this may indicate that it is outside the main group.

The next model is also far more accomplished in terms of the use of space and finishing; it also contains a phenomenon similar to the hand on the previously mentioned piece (Aldred 1973, Cat. 38). A trial section appears on what was clearly an official model of Akhenaten now housed in the Royal Scottish Museum, Edinburgh (Aldred 1973, 98 Cat. 12), perhaps confirming its use as such. The surface of the 4.3 cm thick plaque, measuring 26.3 × 21.3 cm, is filled with a profile of the king, who wears the Blue Crown. The image is finished in detail, with the decoration of the crown, the uraeus, fillet and even flesh-lines on the neck complete. To the viewer's right is a hole pierced through the stone, perhaps to hang or secure the plaque, and beneath the chin of the subject are two attempts to replicate the concentric circles of the Blue Crown, where presumably the copyist has practiced the circle decoration before committing it to the crown. Furthermore, the process is illustrated in a second attempt that shows the central

188   *Sally-Ann Ashton*

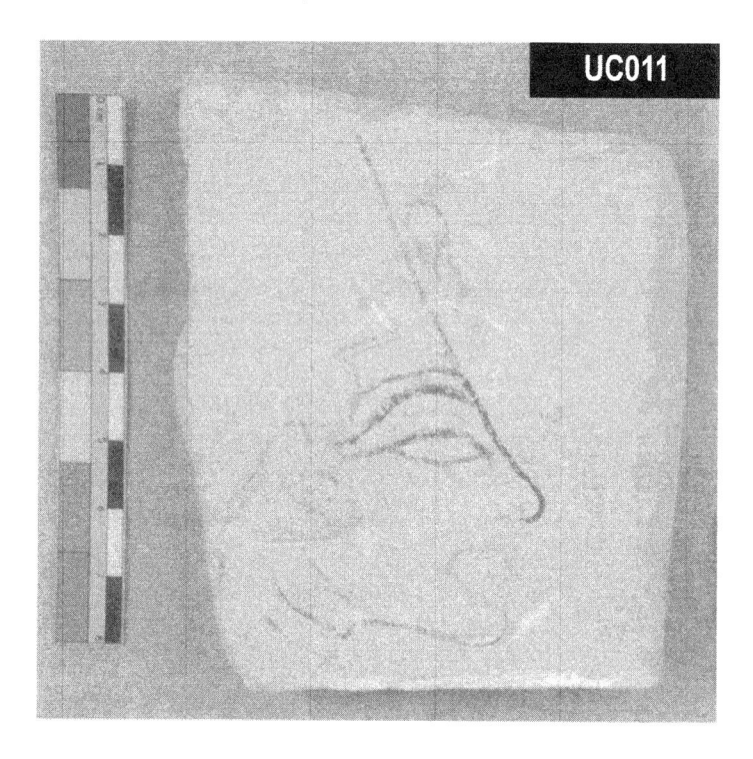

Figure 9.8 UC 011 (Copyright and reproduction courtesy of The Petrie Museum of Egyptian Archaeology, London).

point and the outer circle. The model has a find-spot, which again illustrates how such models might be discarded. It was found by Carter and Petrie at Tell el-Amarna, in the vicinity of stela X. Aldred suggests that it was used as a model for the boundary stelae on account of the Blue Crown, which is not commonly found on relief representations of the king. The features are less exaggerated on this piece than on many of the early period Amarna sculptures and reliefs, or perhaps appear to be on account of the skillful carving. The features are similar to those of the head sketched on UC 011.

UC 011 (Figure 9.8) is worthy of further consideration (Aldred 1973, 59). The subject seems to wear the tall headdress and so is likely to be Nefertiti, although the Blue Crown cannot be dismissed without doubt, because the small rounded side flap in front of the ear occurs on both forms of crown and is indeed more prominent on the Blue Crown. Although it is difficult to see how this piece could have functioned as a model, because it is only partly complete, it could have been used as part of a teaching tool (Pavlov 1941), showing how the image is drawn and then carved, or we must conclude that it was started but abandoned for whatever reason. It is also possible that it was necessary to have a reminder of the form of the mouth in relation to the 'portrait' form. Without a greater under-

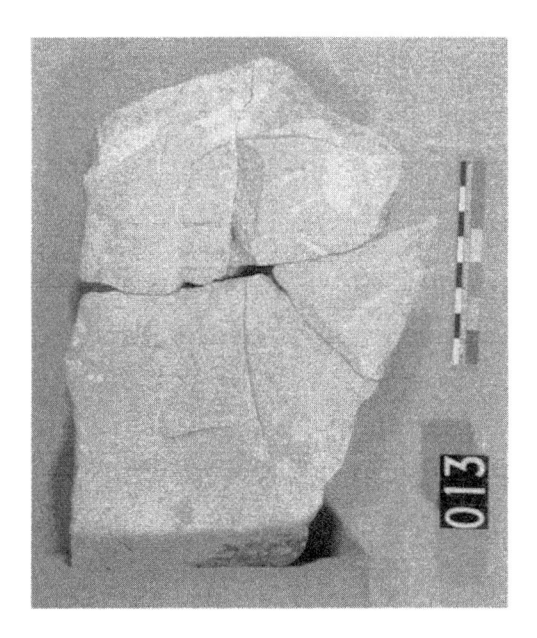

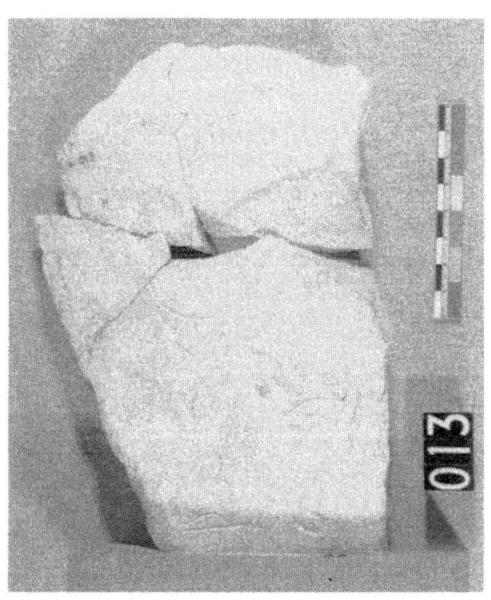

Figure 9.9-10 UC 013 (Copyright and reproduction courtesy of The Petrie Museum of Egyptian Archaeology, London).

standing of sculptors' practices it is impossible to know. This example could, in theory at least, have functioned as either trial or model. However, a close inspection of the piece reveals a high standard of craftsmanship and carefully executed borders to the relief carving, similar in fact to those around the back of the *Afnet* Crown on Aldred's Cat. 15, giving an almost three-dimensional appearance to the relief. This piece is roughly finished at the back and only the right edge is preserved. In terms of placement this is typical of the formal models, but the thickness of the plaque is only 1.6 cm, suggesting perhaps that the back was damaged either in antiquity or later. If this did function as a trial piece then it is likely to have been one for the master sculptor, who for whatever reason gave up on the piece. This raises another interesting possible function, that of trial official image; a sort of practice piece for the master sculptor, which would then presumably be approved or cleared by the royal family or advisors to the king. The fact that this piece has in part been sketched in black ink distinguishes it from the other trial pieces, and the quality of carving of the mouth supports this interpretation. A similarly executed (sketched) plaque but more complex scene is illustrated by the model of a princess eating a duck (Cairo, Egyptian Museum JE 48035, Arnold 110–12, 135, Cat. 44).

Another well executed but partially finished model that was re-used in antiquity is also now housed at the Petrie Museum (UC 013, Figures 9.9–10). The piece is in four rejoined pieces; none of the original edges survive. The sketchily executed head has swollen but rounded lips forming a down-turned mouth and almond-shaped eyes, typical of the later Amarna period. The headdress probably shows the start of an *Afnet* crown, rather than the Nubian wig, which is more common on the early representations. The lower section is unfinished and so it is difficult to see for sure. The top of the head is rounded, but the breaks prevent the viewer from seeing if the more modelled *Afnet* crown was scratched onto the surface. Interestingly the back of this model was re-used, upside down to the main, single figure. On the reverse are two eyes and three mouths, all stylistically similar to those of the main figure. There are also two *nb*-signs and a sketch probably representing a female figure. The fact that the sketch wears a different crown to the main figure would suggest that the block was re-used as a trial piece rather than the back simply being used as a practice surface for the sculptor using the model.

A piece stylistically similar to the Scottish model was found in the temple of Aten (JE 59294). This example shows two heads of slightly different styles. It has been suggested that one head is Akhenaten and the other is Smenkhkare. Both subjects wear the *Afnet* headdress and have an unfinished uraeus. The necks are shown at different angles. The portrait features are also different, with the left-hand example showing a slacker jaw line and open mouth; it appears that the right hand image was carved first, on account of the negative space to the immediate proper right. The reason for the squashed appearance of the headdress on the proper left figure seems to be lack of space. Aldred suggests that this is a model on account of the quality, a theory with which I would agree because a survey of representations from this period illustrates differences in facial features; and so it would be necessary to have models with these variations. Whether or not it represents two rulers is questionable, in that they both face the same way. This would suggest that, rather than an interactive relief model for temple or stela decoration, the Cairo piece is a trial. The negative space below the necks confirms this. It is extremely important to consider the function of these double portraits.

A second plaque, now in the Brooklyn Museum of Art, is later in terms of its more naturalistic style of portrait, and is more like an earlier plaque (Aldred 1973, Cat. 38) as it has two (in this case quite distinct) subjects who face each other. Both Nefertiti and Akhenaten appear on these pieces; on the Brooklyn example, bought at Amarna, the royal pair appear in profile, facing each other (Aldred 1973, Cat 121). Aldred categorized these two pieces, along with another (Aldred 1973, Cat. 115) as models on account of their complete state and also the quality of carving. The Brooklyn plaque has traces of red paint on the surface, which would support Aldred's interpretation that the piece functioned as a model. However, some scholars have suggested that this image is symbolic in that it represents the co-regency of Akhenaten and Nefertiti (Allen, 1991, 74–85; Murnane 1995, 205–8). Such an interpretation suggests that these so-called 'models' served a very different function and I think it possible that the appearance of the two rulers can be explained as part of the increase in the appearance of the image of Nefertiti in Aldred's middle and late periods, as previously noted with regard to the individual relief plaques. It is, however, interesting to note that there is an almost exact equivalent found in the Ptolemaic period, whereby an early Cleopatra is shown with her son, probably as regent or certainly during the time that either Cleopatra I or III acted as regent (Ashton in Walker and Higgs 2001, 68 Cat. 47). On both pieces the intentional duality is confirmed by the fact that the two are facing each other, a feature that has important ramifications for the previously discussed so-called student piece, but possible model (Aldred 1973, Cat. 38).

There is a further link between the Ptolemaic and Amarna periods in the case of Aldred's Cat. 119. There is a portrait of Akhenaten on this late Amarna period plaque, defined as a practice piece by Aldred because it is unfinished, and on the reverse is the outline of the hieroglyphic sign of an owl. The piece was found at Amarna by Petrie, and is thought to have come from a sculptors' workshop north of the palace. The relief is broken on all sides, and it would seem, given that it was re-used, that the original portrait functioned as a model, contrary to Aldred's conclusions. The appearance of the owl sign is of particular interest because it also adorns many of the Ptolemaic plaques. Its suggested provenance as a sculptors' workshop north of the main palace, which might explain why it was re-used.

Aldred, however, described a second example from the late Amarna period as a master sculptor's model (Aldred 1973, 185 Cat. 115). Like the other plaques that I have designated in the category of formal model, this piece fills the available surface space and shows the neck and head. The uraeus is unfinished, once again suggesting that only parts of the image were shown (perhaps only those features essential for the sculptor carving the new style). However, there is no trace of colour on the surface. This particular piece, now housed in the Egyptian Museum, Berlin was found during excavations in 1914, in House O, north of the main Wadi.

Petrie Museum UC 402 (Figure 9.11) is also a formal plaque dating to the late Amarna period. The upper section is missing, but the lower section shows the head of Nefertiti wearing her tall headdress. The execution is of the highest standard, as seen by the detailed carving of the ear, which presents an almost three-dimensional quality. This is also true of the variation of depth of carving on the chin, nose and mouth. Details such as the neck lines are more naturalistically placed and carved than many, particularly early, relief images. In fact, this piece could easily have served as a master model to illustrate the various techniques of depth. In more ways than one, this plaque is an equivalent to the more famous bust of Nefertiti and probably served as a model. The back is roughly finished but there is no evidence of re-use.

Finally, a stylistically similar piece also now in the Petrie Museum (UC 036, Figure 9.12) illustrates a unique form of re-use. There are two faces on one side, and the back has been prepared and carved as if for a figure. The back shows a proper right arm by the side of a torso, and the left arm was probably held out, although it is now clear what the sculptor intended below the elbow. Clear tool marks survive, but the actual composition is unclear. There are also two lines cut to form a rough dividing line between the arms and the body and the entire

192   Sally-Ann Ashton

image is upside down if considered alongside the main image. The same is true of the sketchier head underneath what appears to be a perfectly respectable but sketchy profile of Akhenaten. Traces of pigment survive in the cut around the chin of the more formal and larger head, suggesting that the block was re-used. The smaller head beneath what we can assume to be the larger original shows a subject wearing the *Khet* wig. This re-used block is similar to many of the trial pieces that simply re-use space on both surfaces, but whether the larger head was an abandoned or partially finished model, is difficult, in this instance, to say.

   Although Aldred differentiated between model and trial piece, at face value it seems that his reasons for the differences between the two sub-categories are rather tenuous. Clearly some models at least were re-used and some were left unfinished. The important point here, however, is that whilst there are doubtless examples of trial pieces, did they take the same form as the official models? This is not easily answered, because we are forced to judge the skill of an artist in order to determine if a relief was carved at the hand of the master or pupil. It is possible that some of the less carefully positioned relief images form early models, and that during the early years of Akhenaten's reign a more formal form was introduced and produced. Many examples from the relief plaques can be

Figure 9.11 UC 402 (Copyright and reproduction courtesy of The Petrie Museum of Egyptian Archaeology, London).

Egyptian Sculptors' Models 193

associated with the dissemination of the royal image, and perhaps more importantly there is evidence for the re-use, that is to say re-carving, of models. Both points represent a great step forward when we consider that scholars cannot agree over whether the Late Period/Ptolemaic parallels of the Amarna relief models were concerned with manufacture or were dedicated as ex-votos.

It seems therefore possible that the relief representations functioned as models for trained sculptors and that at times the artists would have used the back or even front as a test area before carving onto the temple wall or stela. In this sense the plaques could have functioned as 'trial pieces,' but not in the sense that the sculptors needed training. I would prefer here to use the term 'working model,' if

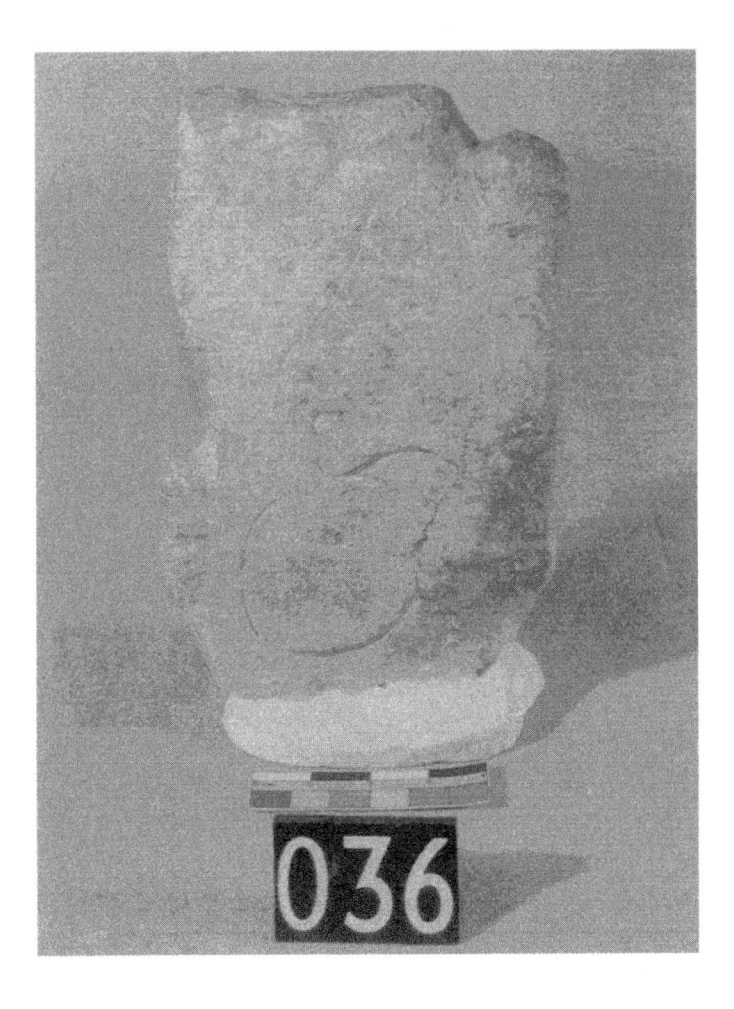

Figure 9.12 036 (Copyright and reproduction courtesy of The Petrie Museum of Egyptian Archaeology, London).

194 Sally-Ann Ashton

we can agree that the objects functioned primarily as models made by the chief sculptor or chief of works for those in his workshop. The models may also have functioned as trial pieces for the chief sculptor as illustrated by the partially sketched UC 011.

With regard to the relief plaques it is worth noting the relevance of the continued production of models beyond the initial most dramatic artistic changes. The first difference to note is that in the middle and later periods Nefertiti's image is more common to the repertoire; secondly, there seem to be more models in the early period, which may be relevant when attempting to understand their function. If they did function as models then one would expect more in the period immediately after the introduction of artistic innovation. Because of the developing styles of the Amarna period we would expect new models for each phase, and consequently, there are models that date to the later phases of royal representation. During other periods there are sporadic examples, but these mostly appear to be trial pieces on re-used blocks of stone. The vast majority of formal relief plaques date to the Ptolemaic and Amarna periods. However, we

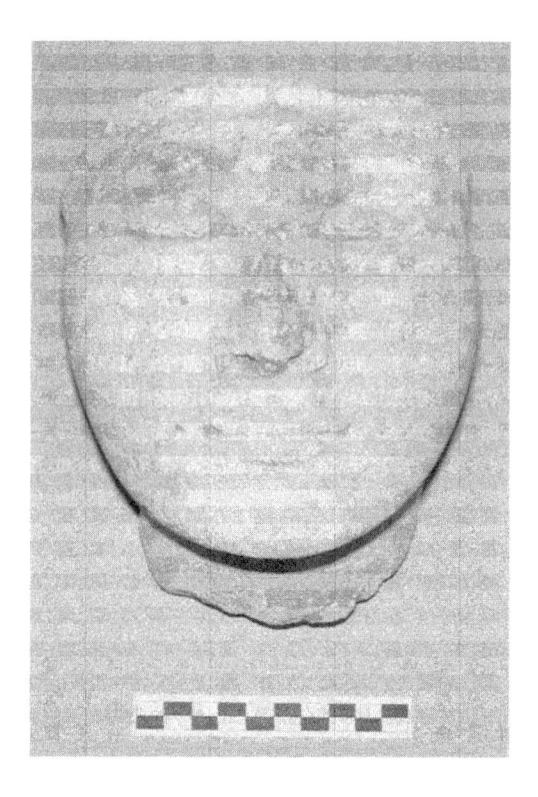

Figure 9.13 EGA.4696.1943 (Copyright and reproduction courtesy of The Fitzwilliam Museum, Cambridge).

know from demotic inscriptions that some of the Ptolemaic examples were indeed used as ex-votos (Bothmer 1953; Young, 1964). However, there is no evidence to support this function during the Amarna Period, and quite the contrary is true.

MODELS IN THE ROUND

A closer consideration of known workshops excavated at Amarna is necessary in order to substantiate these initial conclusions regarding the relief plaques. As with the first group, there are post-18th Dynasty examples of plaster models, as illustrated here by the Third Intermediate Period royal portrait, now in the Fitzwilliam Museum (EGA.4696.1943, Figure 9.13). This form of plaster head accords with the material from the Tell el-Amarna workshop of the sculptor Thutmose.

Borchardt found the workshop of Thutmose during excavations at Tell el-Amarna in 1912, and, aside from providing us with arguably the best-known icon of ancient Egypt, it is extremely important for our understanding of the working methods of sculptors. The material provides us with a second category of potential models in the form of sculpture in the round, in both plaster and limestone. More importantly, however, it illustrates the relationship between Thutmose who was 'Chief of Works' as well as sculptor (Arnold 1996, 41). As Chief of Works, it is possible that Thutmose was in charge of the royal image, and such a position would require a certain amount of collaboration with the king. If we consider the relief plaques in this context it is easy to see how pieces may have been presented to the ruler and adjustments made according to preference. Arnold suggested a similar scenario with regard to the limestone heads that were found in the workshops of Thutmose and I would suggest the relief plaques should also be viewed in a similar light.

It is within this context that the bust of Nefertiti has been placed (Arnold 1996, 65–69, for discussion and earlier bibliography). Krauss has illustrated the grid unit that results in a more carefully proportioned image and it is for this reason that Arnold suggests the piece functioned as a model for other works. The piece is finished in every sense. Plaster covers parts of the limestone core and the surface is painted. In support of the idea that this piece was a model is the missing inlay from the eye, which was not found at the site in spite of the promise of a reward by the excavator to those working there. An inspection of the inside of the socket has also revealed no traces of bonding agent. Furthermore, as Arnold points out, there is no evidence to support the suggestion that a bust could function as a cult object in private houses and this is corroborated by later busts from the Ptolemaic period. The issue of the original inclusion of a second eye, discussed by Arnold at length, does not affect her arguments and, to be realistic, its presence would be expected given the careful attention to detail in the modeling of the piece and particularly given the overall finished appearance.

This particular form of model bust is only found in the Amarna and Ptolemaic periods. In the cases of the Ptolemaic examples, they seem to have been used to continue the styles of the 30th Dynasty rulers. In the Amarna workshop the models appear to have been used as the definitive image for a new artistic style, in this case the later style of the Amarna period. The rationale behind the two functions is different but the general use is the same. More importantly here, the lack of examples from other periods may offer support for an innovation in the working methods of sculptors.

A second unfinished head of Nefertiti with ink still visible on the brows and eyes was also found in the workshop (Aldred 1973, Cat. 100; Arnold 1996, 66 Fig. 61). This head was found during the 1912–1913 excavations in house P 47.1–3, the workshop of Thutmose. The piece is preserved to the neck and unlike composite pieces there is no extension for slotting the piece into the body; the long neck is standard for the finished images of Nefertiti. Unlike the bust of Nefertiti there is no base formed from the shoulders and so it is difficult to see how the piece could have stood unaided. Aldred described this piece as an unfinished head of the queen, whereas Arnold includes this piece with the bust as a model, albeit unfinished (Arnold 1996, 69–70). Given that the head is of the same form and presumably date as the bust, it is difficult to see why there would be a need for a second model, unless it was intended for another workshop. The traces of pigment on the brows, eyes and neck and in particular the shading on the subject's left cheek may indicate that this was an early prototype, carved by Thutmose or one of the chief sculptors, and that the more famous bust was the finished product and model. Once again we have the possibility of a master sculptor's trial piece in this head.

The other important find from the workshop is a series of plaster or gypsum heads of both royal and private individuals (Aldred 1973, 107–112; Arnold 1996, 46–51). The potential survival of such pieces is not good and examples from a workshop at Ptolemaic and early Roman Memphis illustrate their fragility. Plaster heads were used in a variety of functions, not least in funerary context from the Third Intermediate Period, as illustrated by finds from Memphis and Saqqara (Ashton 2003, 60–64). Roeder in his 1941 publication showed that heads of similar appearance were used during the process of manufacturing sculpture and the sporadic examples from other periods, mainly the Third Intermediate to Ptolemaic periods, show that their use was not restricted to periods of artistic change. The hollowed backs of these images and the finished appearance of the faces suggest that they were taken from images, perhaps for checking with the client. Arnold has convincingly explained the lack of stone equivalents found in the workshop with the fact that the finished pieces would not stay in the workshop. She has also made associations between sculptures and plaster heads. The plaster examples were collected and sealed within a room before the occupants left the house/workshop. Had they not been abandoned it seems likely that with no further use they would have been destroyed by the artists, which might explain why there relatively-few survive from other periods of Egyptian history. Casts of hands and body parts such as Fitzwilliam Museum, EGA. 4517.1943 (Figure 9.14) may also have been used as part of the general process of manufacture.

The plaster heads from the workshop then, are probably not the result of artistic innovation, but rather indicate the methods used by sculptors, modeling

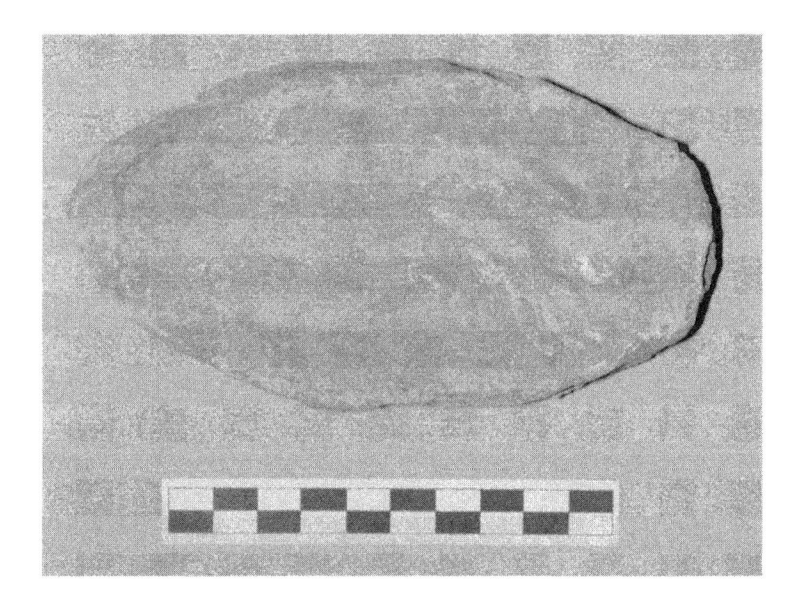

Figure 9.14 EGA.4517.1943 (Copyright and reproduction courtesy of The Fitzwilliam Museum, Cambridge).

in clay and taking a mould in plaster to allow a truer likeness of what would be the finished piece. The plaster heads are therefore models in the truest sense. The bust of Nefertiti could be a model as suggested by Arnold and one would expect a definitive image to be held within a workshop for reference, perhaps carved by the chief sculptor for those under his control. With regard to the plaques, to suggest that these junior sculptors serving the king need basic lessons in carving seems somewhat improbable when we consider that the royal house had a choice of trained master sculptors from throughout Egypt. Indeed it has been suggested that the Tell el-Amarna sculptors came from Memphis or Hermopolis, on account of the more fluid images compared to the early Karnak sculptures (Arnold 1996, 22). It seems that what we might be dealing with in the Amarna period is simply a better stock of evidence for practices that must have ensured continuity in the style of the royal image. Similarly, Petrie's excavations at Memphis uncovered a workshop for which we have less archaeological evidence. Tempting though it is to assume that lack of evidence means a change in artistic practice during these innovative periods, caution should be exercised before concluding that models were only used at such times. It is fair to say that surviving examples show a desire to control the royal image, but how do we account for the comprehensive styles that can be found throughout Egyptian history? The Amarna and Memphis workshops may just answer that question.

So, what conclusions can we draw? Both the Ptolemaic and Amarna periods are times when the royal image was carefully controlled and when there were clear innovations in the representation of the ruler, and an increase in the representations of the royal women. What we need to establish is how closely reliefs and statues compare when they are considered within geographical areas. If the same group of sculptors were employed, and traveled around, then there would have been less need for models to disseminate the royal image because the master sculptor would control this wherever he was. Models would be needed for the Amarna and Ptolemaic periods for two different reasons, either on account of a new style, or because of the continuation of an existing style. In both cases there seems to have been a close link with the royal house.

The evidence from other periods is important in drawing conclusions in this respect because if we find exact replicas of the royal image throughout Egypt then models must have been employed. What is needed is a general art-historical survey of reliefs and sculptures in order to answer this question, and more research in the form of the 1987 conference papers on the artistic presentation of Amenhotep III (Berman 1990).

ACKNOWLEDGEMENTS

I would like to thank Janine Bourriau and Jacke Phillips for inviting me to present this paper and for their comments on the text, and Rachel Sparks for her response to the original paper. I am also grateful to Lucilla Burn and Helen Strudwick for their comments on the text; any errors remain my own. Finally, I would like to thank Hugh Kilmister for his help in obtaining images of the Petrie Museum material and the Petrie Museum of Egyptian Archaeology for giving permission to reproduce images of their material.

REFERENCES

Aldred, C., 1973, *Akhenaten and Nefertiti*. The Brooklyn Museum, Brooklyn.
Allen, J., 1991, Akhenaten's 'Mystery' co-regent and successor, *Amarna Letters*, 1, 74–85.
Arnold, D., 1996, *The Royal Women of Amarna: Images of Beauty from Ancient Egypt*. New York, The Metropolitan Museum of Art.
Ashton, S.-A., 2001, *Ptolemaic Royal Sculpture From Egypt: The Interaction Between Greek and Egyptian Traditions*. Archaeopress, Oxford.
Ashton, S.-A., forthcoming, The double and triple uraeus and Egyptian royal women, in A. Cooke and F. Simpson (eds.), *Current Research in Egyptology 2001*. Archaeopress, Oxford.
Ashton, S.-A., 2003, *Petrie's Ptolemaic and Roman Memphis*. Institute of Archaeology, London.
Berman, L. M., 1990, *The Art of Amenhotep III: Art Historical Analysis*. Cleveland Museum of Art, Cleveland.
Bianchi, R. S., 1979, Ex-votos of Dynasty XXVI, *Mitteilungen des Deutchen Archäologischen Instituts, Abteilung Kairo*, 35, 15–22.
Bianchi, R. S., 1980, Not the Isis Knot, *Bulletin of the Egyptological Seminar*, 2, 9–31.
Bothmer, B. V., 1953, Ptolemaic reliefs IV. A votive tablet, *Bulletin of the Museum of Fine*

Arts, 51, 80–84.

Edgar, C.C., 1906, *Sculptors' Studies (Catalogue Général)*. Cairo Museum, Cairo.

Josephson, J. A., 1997, *Egyptian Royal Sculpture of the Late Period 400–246 BC*. Von Zabern, Mainz.

Krauss, R., 1991, Nefertiti- a drawing board beauty? The most lifelike work of Egyptian art is simply the embodiment of numerical order, *Amarna Letters*, 1, 46–49.

Lauffray, J., 1979, *Karnak d'Égypte. Domain du divin*. Éditions du Centre National de la Recherches Scientifique, Paris.

Lauffray, J., et al., 1970, Rapport sur les travaux de Karnak, *Kêmi*, 71, 57–99.

Murnane, W. J., 1995, *Texts from the Amarna Period in Egypt*. Scholars Press, Atlanta.

Myśliwiec, C., 1972, Towards a definition of the 'Sculptors' Model' in Egyptian Art, *Études et Travaux* 6, 71–5.

Pavlov, V. V., 1941, About the sculptors' models in Egyptian art, (in Russian) *Iskusstvo*, 5–6, 67–70.

Roeder, G., 1941, Lebensgrosse Tonmodelle aus einer altägyptischen Bildhauerwerkstatt, *Jahrbuch der preussischen Kunstsammlungen*, 62.4, 145–70.

Walker, S. and Higgs, P., 2001, *Cleopatra of Egypt: From History to Myth*. British Museum Press, London.

Young, E., 1964, Sculptors' Models or Votives? In Defence of a Scholarly Tradition, *Bulletin of the Museum of Fine Arts*, 22, 246–256.

Chapter 10

How to Build a Body Without One: composite statues from Amarna

Jacke Phillips

Abstract

The sculpture of the Amarna period has long been familiar as instantly recognisable ancient Egyptian art, and as an anomaly within the overall historical development of that art. One of the most enigmatic aspects of the period is a technique commonly known as 'composite sculpture.' I focus here on certain technical problems in our understanding of ancient Egyptian 'composite' stone statuary, and possible interpretations for the rationale behind their manufacture and for their intended final presentation. This topic is still very much in need of further detailed research, and the paper presented here consists of some observations and tentative conclusions rather than ultimate solutions to the questions raised. This study is ongoing (see Phillips 1994 for an earlier foray), and I raise as many questions as possible conclusions for the phenomenon.

Surviving recognisable examples of 'composite sculpture' have been recovered in the main from the site of Amarna, the new 'virgin' city created under Akhenaten and the site of Egypt's capital from his 6th regnal year to the 3rd year of his successor Tutankhamun (*i.e.*, about 1344–1330 B.C.) when the capital returned to Thebes less than two decades after it had left. Amarna presents us with an unprecedented body of material for investigation, and it is clear that the use of 'composite' stone sculpture was one of the 'new' technologies emphasised during that short period of Egyptian history – but, to our knowledge, one that was not fully realised, and that perhaps ultimately would not have been successful. The evidence consists mainly of unfinished sculptural elements found on the estate of the 'Chief Craftsman and Sculptor' Thutmose (P47.1–3) by Ludwig Borchardt in 1911–1913 (Phillips 1991, 31–40; Arnold 1996, 41–83). The 'composite' technique is often cited as an 'innovation' of the Amarna Period, and occasionally Thutmose is even credited as its 'inventor.' Although he was not the only sculptor at Amarna to use this technique, he is associated most closely with it. Comparatively few 'composite' elements, especially at life-size or large scale, have been recovered elsewhere at Amarna (*e.g.*, Peet and Woolley 1923, 27, Pl. XI.1–2; Pendlebury

1951, Pl. LIX.6–8) and even fewer elsewhere in Egypt, and the Thutmose elements remain absolutely vital to any study of this technique.

Most surviving elements are sculptured heads at life-size scale. Many are considered to represent Nefertiti, Akhenaten's queen, most famous for the limestone bust also recovered in Thutmose's studio and now in Berlin, and there seems little reason to question this attribution. This bust, by the way, is not a finished sculpture, as her 'missing' pupil was never inserted into her left eye. Nor is it a 'composite' element. It is deliberately cut off at the shoulders and upper body, and was never intended to be attached to another sculptural element. Other heads *are* 'composite' elements. They have not been broken from the rest of the finished statue, but rather are, individually, complete in themselves. Attachment of the heads normally works on the 'mortise & tenon' method, as can easily be deduced from viewing the pieces themselves. Most heads have a round tenon at the base of the neck, to be fitted into the mortise hollowed out of a separately made torso element to which it would be attached. On many, the skull is deliberately cut back and prepared with a square or oval tenon of the necessary shape and height to accommodate the separately made crown or wig attachment. The attachment would have been hollowed out underside like the mortise to fit the tenon of the head. The form of the tenons and the cutting of the heads prepared to receive them enable us to identify different types of crowns that these heads were intended to wear. They are a limited group, and have been used by some scholars to identify the person represented. The heads of the princesses, on the other hand, were not intended to be covered by either wig or headpiece, and their shaven skulls are fully carved in the round. A recently recognised example of a finished head is the famous yellow jasper face element usually considered to be Queen Ty, of unknown provenance but probably also from Amarna (Aldred 1973, 107 Cat. 21; Arnold 1996, 35–38, Figs. 27, 29).

These cuttings and tenons, clearly not meant to be visible when the statue was completed, also are found on a number of other body parts as well as the heads – mostly hands, arms, and feet recovered in Thutmose's studio, elsewhere at Amarna, and some elsewhere in Egypt. A wig element in dark greywacke, also recovered in Thutmose's studio, has a squared projection on the interior that would fit onto an equivalent squared depression on the side of a stone head that has not been recovered (Berlin 21272; unpublished). Petrie found another life-size wig element of black granite elsewhere at Amarna in 1892 (UC 076; Samson 1973, 56, Pl. XXVIII; Arnold 1996, 63 Fig. 56). Both consist only of the right half of the wig, and so the head elements for which they were intended equally could have been presented in right profile rather than frontal view, *if* no corresponding left half was in fact produced.

A life-size clenched hand, of red jasper, was found at Thebes near the Dynasty XVIII temple at Medinet Habu (Cairo 59740, Nims 1965, 178, Pl. 89; Hayes 1959, 102; Arnold 1996, 35, 141 n. 91). Hayes also mentions a nose element that actually joins two mouth fragments found by Petrie at Amarna, suggesting it too likely came from there rather than Medinet Habu and also is of Amarna date. Hayes suggested the jasper hand should be dated to the reign of Hatshepsut; this remains uncertain although the solidity and abstraction of the hand and fingers clearly differ from, and surely pre-date, those of the Amarna period. Here, the hand is complete in itself, highly polished where it would be visible on the completed statue, but only roughly finished around the wrist and with a short projection to fit into a corresponding depression at the end of the arm, all intended to be covered from view probably by an added bracelet. A strut along the palm of the hand itself provides further support, and would allow either a vertical or horizontal positioning for either a standing or seated figure. A thumb in the same material now in the MMA, New York, also was found at Medinet Habu, and a heel (UC 150, Pendlebury 1951, 227 #UC 150, III.2:Pl. CVI.5 [misidentified as UC 092]; Samson 1978, 64 Pl. II) and the two mouth elements recovered at Amarna, now in the Petrie Museum, London.

The general interpretation of these sculptures is that each body part is produced in a stone of the correct colour for that specific part, and all the elements discussed so far conform to this interpretation. Yellow, brown or red for hands, feet and heads; black for hair and eyebrows; the eyes made of variety of elements (black, white, quartz crystal), the appropriate colour for whatever crown or even hairstyle was worn; and the torso in white or other suitable colour of appropriate dress. However, black also is known for several heads in obsidian and black granite. In the words of the late Cyril Aldred, "we find feet, hands, arms, and heads in quartzite, granite and jasper for fixing to bodies of white limestone (Aldred 1973, 58). The idea behind this technique presumably was to ensure a more permanent monumental record than a painted statue carved in a single stone, since the individual colours continue throughout the stone itself and could not be worn off over time, as would a painted surface. This technique therefore *seems* to have been a technological innovation in stone used by Akhenaten's sculptors to produce life-size figures of the Pharaoh and his immediate family in some abundance, although at least two statues represented by individual elements in red jasper and obsidian are earlier in date. The technique also seems to have largely died out with the end of the Amarna period.

However, we have no complete 'composite' statue or even statuette, from Amarna or found elsewhere. In fact, we have no separate crowns, no limestone bodies or clothing, no legs and so far only a single fragmentary torso (in red quartzite) at this life-size scale, although we do have a few wigs and smaller crowns for statuettes. The fragmentary near-life-size obsidian elements now recognised as a statue of Amenhotep III, found in the Karnak 'cachette' and of immediately 'pre-Amarna' date (Cairo CG 42101=JE 38248 and Boston MFA 04.1941; Lacovara *et al.* 1996, with further references), probably is the most complete life-size 'composite' statue known. Separate recovered elements are his head and ear, and fragments of his neck (or arm or leg?) and foot. A limestone statuette of Akhenaten now in Cairo (JE 45380, Borchardt 1912, Pls. 1–2; Aldred 1973, 65 Fig. 42; Michalowski 1978, 142 Fig.; Phillips 1994, 68 Fig.) sometimes is cited as a complete 'composite' piece, but the crown does not fit properly, and it is questionable whether the limestone figure and the blue crown were intended to belong together. Even so, only the crown is a separate element and thus the figure should not be considered 'composite' *per se*.

Our rationale and reconstruction of the methodology involved is almost entirely theoretical. It is based almost exclusively on the surviving heads, and occasionally also the few other body parts that rarely are published or studied, as well as extrapolation of the techniques used for large and small wooden statues and for small statuettes in other materials, which are common. But these techniques do not always transfer well to life-size stone statuary. In order to realise the true 'innovation' of the Amarna 'composite' sculpture, and perhaps also to understand just what is happening here, we must look at the earlier technology upon which it draws.

This technique strongly relates to 'composite' sculptures in wood of the Amarna period, such as the small head of Akhenaten's mother Queen Ty in Berlin that employs three different kinds of wood and exhibits at least two different stages of presentation (21834, Aldred 1973, 81, Pl., 105 Cat.19; Wildung 1995; Arnold 1996;30–35). The wig she now wears, made of linen and beads, covers an earlier representation of the queen wearing a silver *khat* headpiece over her head of acacia wood, with gold earrings and uraeus, and other royal accoutrements of various metals and other materials. A tenon protrudes at the bottom of her neck, for attachment to a (missing) body and, at the second stage, she wore a *modius* headdress and a tall plumed crown that was inserted into a second tenon still visible atop her head.

This use of multiple different woods for 'composite' statuary in wood can substantiated back at least as far back as the XIIth Dynasty, when a wooden 'composite' head of a woman and accompanying separate wig was produced (Michalowski 1978, 48 Fig., Salah and Sourouzian 1987, Cat. 89). The woman, found at Lisht, has several small projecting dowels on her head to hold the two pieces of a wig in place, and these are now visible because the top of the wig is damaged. She also has a larger neck tenon for insertion into a body, now missing. At least two different kinds of wood were used in her manufacture, one a light hardwood for the head and the other darker softer wood for her wig with added gold appliqués. Her 'real' hair beneath the wig is painted at the top of her forehead. Contemporary use of stone rather than wood is exemplified by a Dynasty XIII female statuette found at Thebes, her body carved in ivory and her hair separately carved in gypsum, the latter painted and flecked with gold (Reeves and Taylor 1992, 101 Fig. left). Both materials, however, are white in colour.

Wood was a scarce and expensive commodity in ancient Egypt, and virtually all hardwoods useful for producing statuary as well as fine furniture were imported from abroad. Most wooden statues consist almost entirely of smaller pieces of wood for a number of separate individual elements, and the same type of wood need not necessarily be used for the entire statue, although it most often was. As much as possible of every piece of wood would be used, so normally all projecting parts were made of separate pieces. Furniture, on the other hand, could employ several types of wood and often was inlaid in finer woods or other materials.

The forefoot of a wooden statue was separately made and added to the bottom of the leg at the heel, as were the projecting kilt, and the forward leg to the lower body, and the arms – either straight in a single piece, or bent in two pieces – to the shoulder. These were attached either with a dowel holding one piece to the other, or having a projection at the end of one piece fitting into a hole on the other in a 'mortise & tenon' arrangement. The tenon or smaller dowel would often be square or rectangular but sometimes rounded or oval. This is well known, and is clearly seen on numerous examples. Large pieces attached by 'mortise & tenon' such as the arm or leg also may have a smaller dowel or two added to secure the joint. This technique works well because wood is somewhat flexible, fitting tightly into the available mortise space. Square and oval tenons and dowels do not allow the added element to be attached in any position except the one intended, and the two pieces are held in place by tension.

Subsequent repairs to wooden statues often also are made in this manner, but stone statue repairs and corrections involve a different technology. Stone, unlike wood, is not flexible. Repairs to objects in stone usually have the damaged area cut down to attach a new and separately carved replacement, using a sliding 'mortise & tenon,' most often with a dovetail joint to hold the replacement piece in position as can be seen on many repaired statues (*e.g.*, Arnold 1996, 62, 63 Fig. 57). Such repairs can be done any time after the object is finished, but also could be made by the original artisan if a mistake has been made or the stone of inferior quality is found in the area itself during manufacture. The dovetail joint (rarely used for wood sculpture) has the important function of holding the addition in place by resisting gravity, and the addition will always rest on another part of the statue. In some cases, it is arguable whether the correction was made during initial production, or was a later alteration. A new head can be added to an old body by same method, as on a now-headless torso now in the Petrie Museum, London (UC 001, Pendlebury 1951, 226 #UC 001, Pl. CV.9).

Another method was employed on an over-life-size alabaster statue now considered to be Seti I, that has been altered by removing the original hands and adding new ones in a different, tightly clenched, position (Cairo 42139; Nims 1965, Fig. 65; Salah and Sourouzian 1987, Cat. 201). Two holes drilled through the wrist area on both the main statue and on each added hand probably held the attachments together using leather strips. A vertical strut may possibly have been added to the hands to fit into the body at the correct point (as on the red jasper hand) but, if so, this is unclear on the statue as preserved. In any case, the wrist area was then filled in with plaster and finally painted or otherwise covered over to obscure or hide the joint. The drill holes now are clearly visible, as this plaster no longer adheres to the statue. Whether this was a change in the original iconography of the pharaoh during its production, or a later alteration, is impossible to say. Such attachments are examples of repair or replacement of a missing or broken area, and do not constitute 'composite statuary,' although the head of this particular statue has been prepared for a separate crown that was not recovered. Arnold (1996, 83) calls such statues, where only the crown is a separate element, "partly composite." The only life-size example from Amarna is a granodiorite head of Nefertiti from Thutmose's studio (Berlin 21358, Arnold 1996, Figs. 41, 72, 74), and she, like Seti, has an integral back pillar at the neck (but see now addendum). The limestone statuette of Akhenaten with its separate blue crown, mentioned above, also would be considered "partly composite;" he too has a back pillar.

Adding separate body elements together as 'composite' statuary was not done in the same way as for statue repair. Surprisingly enough, the dovetail joint is *not* found on 'composite' elements, nor is dowelling or the use of leather ties. The means of joining many surviving body parts are not so easily explained by the same 'mortise & tenon' method employed in wood, under detailed examination. Consider a statue made of wood, with its arms attached by 'mortise & tenon' at the shoulder and imagine the statue and its arms were made in stone instead. Consider the configurations of the arm and hand, to realise that the techniques used here in wood are problematic for stone figures. And yet it seems that the technology of wooden statuary is the primary inspiration for 'composite' sculpture in stone, rather than the technology involved in single-block stone statuary.

Several of the unpublished or little published 'composite' pieces are due to their fragmentary state. Let us take a short look of such pieces which do not conform to the theoretical technology offered by the 'standard' explanation of 'composite' sculpture, but which suggest instead techniques used for the construction of statuary in wood:

Berlin 20495 is a half-life-size right arm made of red quartzite, found in Thutmose's studio, with the arm extending up to about midway along its upper half, and the hand open but slightly curled (Figure 10.1; Borchardt 1912, 34 Fig. 24). The arm is straight and has only one tenon, circular, at its upper end, presumably for insertion into a body clothed in a costume with a straight-edged sleeve that covers the upper half of the upper arm. It does not have the more practical horizontal tenon, as does wooden statuary. The arm could only have been positioned hanging at the side of a standing statue, either with the thumb facing forward or nearest the body. Alternative positions are unlikely. As it has no strut along the arm's length (like the red jasper hand), it would not have been

Figure 10.1 Berlin 20495, half-life-size right arm with tenon at upper end, red quartzite (Borchardt 1912, 34 Fig. 24).

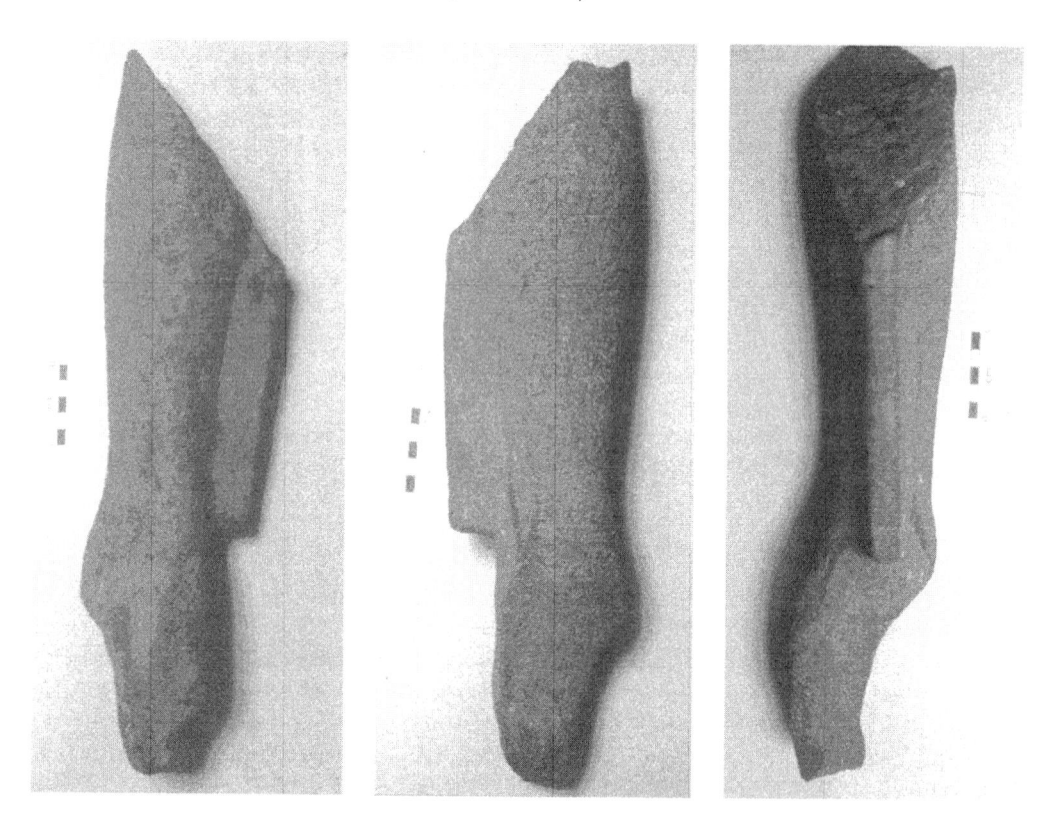

Figure 10.2 (left) Berlin 21244, half-life-size lower right arm and hand, red quartzite. Front view, showing strut.
Figure 10.3 (middle) Berlin 21244, half-life-size lower right arm and hand, red quartzite. Back view.
Figure 10.4 (right) Berlin 21244, half-life-size lower right arm and hand, red quartzite. Side view, showing roughly modelled curled hand.

positioned resting on the leg of a seated statue. It also would not have supported an offering table, as do some life-size statues. In either event, if the arm had been positioned horizontally it would be bent at the elbow.

But, the only tenon would be inserted into a mortise *below* the dress 'sleeve,' not horizontally but vertically. This arm would have been attached *only* from the top, with all the weight of the arm unsupported except at the tenon above – quite a radical venture for a heavy piece of carved stone. The adhesive to be used must have been extremely strong, and the artisan absolutely certain it would not fall out and be damaged. Egyptian stone statues in fact *never* have an unsupported arm handing vertically. Even when an arm is carved in a single piece together with the rest of the body, it is supported by leaving part of the stone linking it to the body or background of the statue. Even separate wooden arms are supported by a horizontal, not vertical, 'mortise & tenon' arrangement. This arm is an aberration in Egyptian art.

Berlin 21244 is a half-life-size lower right arm and open slightly curled hand, made of red quartzite (Figures 10.2–4; Phillips 1994, 70 Fig. right). It has a vertical square-sectioned strut intended to fit into a mortise along the side of the figure at the hip, with the back of the hand forward and the thumb nearest the body, as the palm of the hand has a less polished surface. The figure would be in a submissive pose, possibly bending forward slightly, and hands at its side. The top of the piece is cut off diagonally at about the elbow, presumably for the white limestone of the figure's clothing. This diagonal cut-off would represent a sleeve that extends beyond the arm to the right, and on the left is pulled towards the waist area. This piece, however, has no tenon at its upper end for attachment to the dress above, but only the one vertical strut to hold it in position. The strut is not dovetailed, so it could easily fall out of its vertical mortise, away from the standing figure, without a second support or a strong adhesive securing it against the figure's side. This is an unlikely and ill-considered arrangement in stone statuary.

Alternatively, it might be an arm that was bent at the elbow and in front of the body, with the clothing pulled across the upper arm and elbow to be tied in a knot below the breast, as in a small wooden female statuette (Arnold 1996, Fig. 124). Usually, but not universally, something is held in this hand. This pose would explain the angle of the cut-off, as most figures show this angle reversed and above the elbow, and the strut would fit in vertically across the breast. However, the small statuette and other figures in similar pose have the left (not the right) arm in front of the body, and the strut is in the wrong place relative to the hand, which would have palm (not the thumb) nearest the body. This too is an unlikely reconstruction of the completed statue.

Figure 10.5 Berlin 21225, life-size right arm and hand, with attached left hand, red quartzite. Side view, showing strut and break at edge of left hand.

208   *Jacke Phillips*

Berlin 21225 is a life-size right arm and hand, made of red quartzite, only roughly carved, with the entire arm straight and the hand open but slightly curled (Figures 10.5–6; Phillips 1994, 70 Fig. left). It has a vertical square-sectioned strut all along the back of the arm and hand, and also a very short circular tenon at the shoulder end. A second hand – this time clenched – is beside the first hand that can only be the left hand of a second figure, due to the positions of both thumbs. Thus, this can only represent a double statue, with hands touching but *not* clasped, and in fact in two different positions, a presentation unique in Egyptian sculpture.

Again, the back of both hands faces forwards. The tenon at the top is to be inserted upwards, into the shoulder, like the first piece described above (Berlin 20495). However, the long strut is to be inserted *not* into the body of one figure, like the second piece described above (Berlin 21244), but into a background support behind and between the two figures depicted. This is in effect a very large piece of three-dimensional inlay, in very high raised relief, with both figures attached to a background support. One of the pair has his arm exposed, the other only the hand, an arrangement sometimes seen in double statues of Akhenaten and Nefertiti, although Nefertiti's arm is not entirely covered to the wrist. The identities of the two people cannot be ascertained. Another example is a pair of clasping hands, again from a double statue (Berlin 20494; Aldred 1973, 159 Cat. 87) that may have been from a 'composite' statue but, if so, no evidence of this survives on

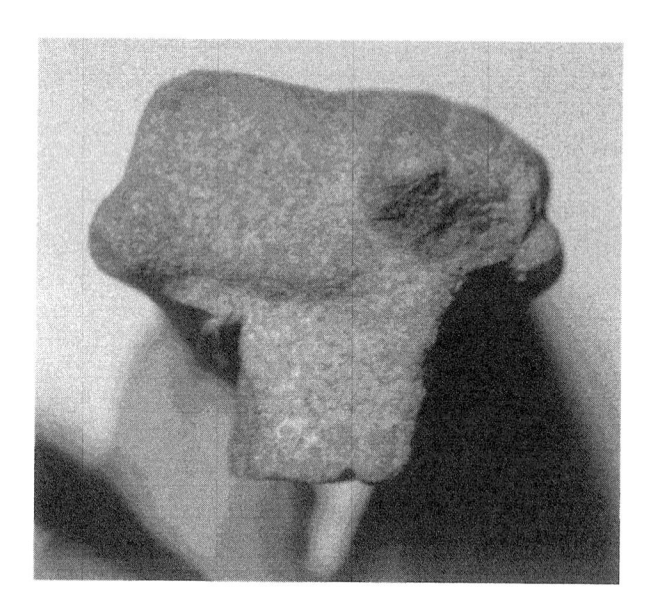

Figure 10.6 Berlin 21225, life-size right arm and hand, with attached left hand, red quartzite. View from below, showing open and clenched hands, and thumbs of both, with strut.

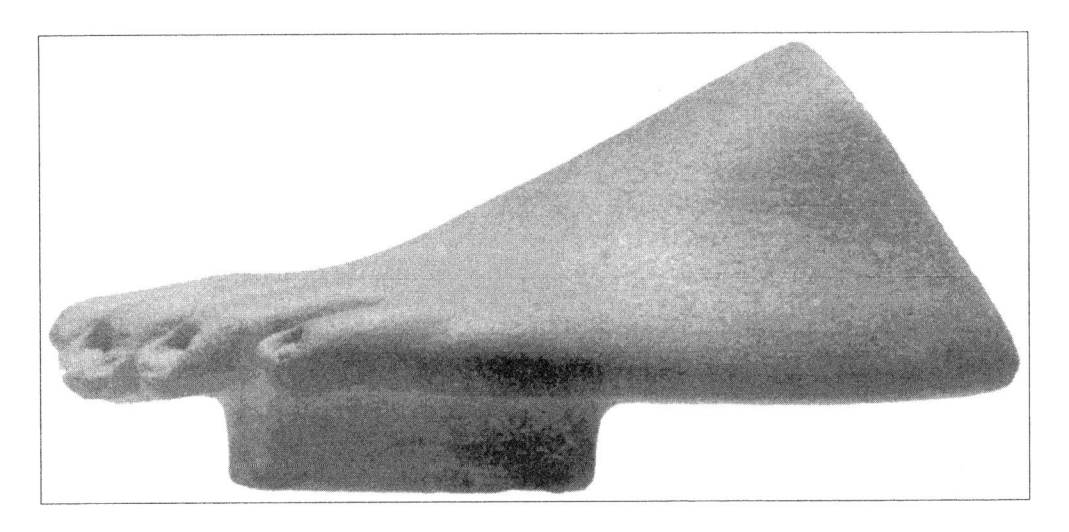

Figure 10.7 Israel Museum 76.14.100, life size forefoot, yellow quartzite (Aldred 1973, 177 Cat.104).

the piece itself. It was found not in Thutmose's studio, but in that of 'Ipy' (P 49.6), the other 'large' sculptor's studio at Amarna excavated by Borchardt (see Phillips 1991, *passim*), and its positioning is exemplified by a statuette of Akhenaten and Nefertiti now in the Louvre (E. 15593; Aldred 1973, 63 Figs. 39–40).

All would have to have been held in place by an adhesive of some kind. Several possible adhesives could have been used, but the most likely is a resin mixed with 'plaster' (*i.e.*, powdered limestone), tinted with a powder of the relevant colour. The relative strengths of ancient adhesives have not yet been tested, to my knowledge, but resin was used for large and heavy stone objects like sarcophagi and vessels (Lucas 1962, 7–8; Newman and Serpico 2000, *passim*), as well as statue repairs. Nonetheless, those pieces were positioned so that such objects still employed gravity to hold the added element together and in position. Yet gravity appears not to have been considered in the production and positioning of all three of these 'composite' arm elements.

Let us now turn to the other limbs – legs and feet. No legs are known for this period, but foot elements from Amarna reveal further problems. Two different and unrelated forefoot elements are known, both representing the left foot and both of yellow quartzite. One is in Jerusalem (Figure 10.7; Israel Museum 76.14.100; Aldred 1973,177 Cat.104), the other in Berlin (21237; Borchardt 1913, 39 Fig. 17). I know both only from their publications. The Jerusalem forefoot has cavities for inlay of toenails in a separate material that need not necessarily be of stone. The other does not, and likely is unfinished. Both are deliberately cut off diagonally at the mid-step, presumably for attachment to the hem of a dress likely of limestone or, less likely, a leg or heel also in yellow quartzite. Yet neither indicates a tenon, mortise or any other means of attachment at this mid-step junction, which is flat and smooth. We may presume both forefeet are intended for female statues, since they are the yellow colour that normally indicates female skin in Egyptian convention, and so may have been intended to 'peek' out from under a floor-length dress. A long horizontal strut can be seen on the sole of the Jerusalem foot for insertion into the statue base, but so such strut is shown on the other. Both feet would need to be inserted to the statue horizontally, due to the diagonal mid-step, and could not have been intended to actually *support* or help to balance the rest of the statue above for the same reason.

The Jerusalem forefoot also has indications that it had worn a sandal, according to the published catalogue description, as traces of gold were found along the visible edge and two drill holes are located between the first two toes. A sandal fastening (Berlin 30317 + 30344-30347; unpublished) composed of several separate elements of travertine, also was recovered at Amarna and shows how they might have been arranged together on a 'composite' foot. This is a typical sandal fastening of the period, in the appropriate white colour, and also is shown on Amarna frescos (*e.g.*, Ashmolean 1893.1–41 [267]; Arnold 1996, 57 Fig. 49), but the Berlin fastening is not associated with either left foot discussed here; it was recovered in 'Ipy's' studio (P 49.6), whilst they were not.

Both forefeet seem to serve as what can only be described as a 'wedge' to support the much larger intended (limestone? or travertine?) costume above, but neither would support the much larger dress element firmly as it is not mortised in at the junction and, even with the tenon below on the Jerusalem forefoot, would do little to keep a large otherwise unsupported element such as a dress from falling. Thus, in all likelihood, the dress, its back support and the base on which either forefoot was attached would all need to have been a single piece of stone for the final resulting figure to be viable as a standing statue, although the arrangement might have worked for a seated statue. The only torso element known includes an integral back pillar (see Phillips 1994, 66), and no seated statues have been suggested from known 'composite' elements. The Jerusalem forefoot is a large and elaborate inlay element for a very high raised relief figure. The Berlin left forefoot element, lacking both strut and mortise or tenon, emphasises this conclusion, since it has no means of attachment to any other statue element at all.

These pieces all are problematic, if we reconstruct the entire finished figure or figures as free-standing statuary, but they make much more sense as large scale inlay elements for life-size or half-life-size high relief figures that are integral with a vertical backing support. They may be compared to the high relief figures of the Amarna boundary stelae, consisting of Akhenaten, his wife and two or more of their daughters. Those, however, are carved in one piece from living rock, with no inlay work.

If the 'composite' elements were combined, they should have produced something similar to a standing statuette of Akhenaten and Nefertiti, now in the Louvre (E. 15593; Aldred 1973, 63 Figs. 39–40), or the Aswan stela of the sculptor Bak and his wife now in Berlin (Figure 10.8; Berlin 1/63; Aldred 1973, Fig. 6).

Figure 10.8 Berlin 1/63, stela of the Sculptor Bak and his wife (Aldred 1973, Fig. 6).

Akhenaten's and Bak's left arm both are similar to Berlin 21244, the long vertical arm shown before with the back of the hand forward and Nefertiti's forefeet are similar to both those just discussed above, her dress hiding the back half of her feet. Both statuette and stela are in high raised relief, but are not 'composite,' being carved from a single piece of stone. For the 'composite' heads, the back of the head is as finished as the rest of the element, so it is unlikely that any supporting back pillar would have continued any higher than the shoulder, with the attached head and its added covering entirely in the round. Yet the back pillar in Egyptian art continues at least to midway up the back of the head, high enough to support the head as well as the body; this too can be seen on the double statuette (as well as those only "partly composite"). The theoretical back pillar rising only to shoulder height – if it existed – also would be a feature of 'composite' statues otherwise unprecedented in Egyptian sculpture. The granodiorite head of Nefertiti (Berlin 21358) the only "partly composite" life-size head found at Amarna integral with her body, includes a back pillar at her neck.

The most puzzling piece of all is a life-size heel element made of red jasper from Amarna, now in the Petrie Museum (UC 150; mentioned above). It is extremely well-made, highly polished on all surfaces, but with no means of attachment anywhere. A similar but unpublished element, made of yellow quartzite and now in Berlin (24413), has a tenon on the sole like the yellow quartzite forefoot but, like the jasper heel, none either to the forefoot in front or to the clothing (or possibly leg) above. We might also wonder why the foot element constituted only the heel, since no known footwear at that time would account for completely covering only the forefoot. It may be that the foot normally was made in two pieces, its junction hidden by the sandal strap; this was common on the forward foot of wooden statuary. If so, the two elements would be far less stable than a single piece and, in any event, no red jasper forefoot has been recovered. If anyone can think of an explanation for this jasper heel and how it was intended to function, I would be very interested.

We are still faced with the problem that torsos (barring the one mentioned above, with a back-pillar) and clothing elements for these heads, arms and feet have not been recovered – and that the number of unfinished and even some finished heads are far too numerous for the quantity of arms and feet that do exist. Most likely these elements, recovered we must remember in a sculptor's studio and mostly in an unfinished state, probably represent unfinished or incomplete compositions. The body/back/base might have been produced elsewhere, perhaps at another sculptor's studio not yet excavated. But the planned life-size pairs and individuals were never actually completed before Akhenaten died, and the project would have been then abandoned in the troubled times that followed, before the city itself was abandoned in Year 3 of Tutankhamun. But, even so, not all elements can be explained this way. And, after all this discussion of the elements themselves, how does *our* workshop theme of 'social context' fit into the sudden introduction of the new technology in large-scale stone sculpture at Amarna. I have considered a possible explanation.

The city of Amarna was a vast new complex, constructed from nothing and abandoned by all but squatters within 20 years of its inception. The circumstances of embellishing it are unparalleled in ancient Egypt: construction of a completely new capital city in honour of a new deity, the Aten. The quantity of new stone sculpture required by the king, his new religion, his court, and his élite would have been immense. The leftovers that have been found in excavation so far are indication enough of the huge scale of that production during Akhenaten's reign.

It seems to me a strong probability that artisans from throughout Egypt must have been drafted and brought to Amarna to produce what was required to embellish the new capital as quickly as possible. The situation was unprecedented, and the trained number of sculptors in stone likely was insufficient for Akhenaten's requirements. Trained sculptors may have been drafted to work on material they were unused to using, that is specialist sculptors of wood were drafted to produce sculpture in stone. We know from the ancient records that sculptors did specialise not only in certain types of objects and techniques but also in the types of material employed to make them.

I believe I have shown that the technology behind the pieces found in Thutmose's studio reflects more the methodology of wood sculpture rather than stone, even if the results are not quite the same. It is possible that the 'composite' pieces are in fact a series of failed experiments on the part of Thutmose and his workers, and possibly other establishments elsewhere in the city. This may explain why the individual pieces are so well made and well presented, but the technology implied in the manufacture of the pieces I have introduced (and others I have not) generally is untenable for their scale and material. The arm with vertical tenon at the shoulder end but without struts (Berlin 20495) might be argued as an initial but highly unsuccessful foray in stone by a sculptor normally employing wood, because it would have worked if it had been made of wood. The two arms having struts (Berlin 21225 and 21244) might be considered as later but still unsuccessful second-stage attempts to alleviate the problem by sculptors still not comfortable in an unfamiliar medium.

This might explain why so many well-presented but technically unusable elements have been recovered in a sculptor's studio complex that was abandoned, not destroyed. Virtually all elements recovered here are unfinished. Whilst we may argue that their subject matter no longer was required when the city was abandoned, this would apply only to the heads that recognisably belong to the royal family. The hands, arms, and feet are not distinctive in the same way, and theoretically could have been reused with no implied disrespect to the new order. If these elements were still useable as sculptures, why would they not be removed when the studio was abandoned, together with the city?

ADDENDUM

Recent fieldwork at Amarna has recovered and pieced together multiple fragments of an unfinished seated double-statue of Akhenaten and Nefertiti initially excavated by the Germans at the home of the master-sculptor Thutmose. The team also has conclusively demonstrated that the head of Nefertiti to which it joins is the unfinished 'partly composite' granodiorite head (Berlin 21358) having a back-pillar but without neck tenon discussed above. Thus, this head clearly is distinct from the others, which *are* composite pieces, in that it was indeed integral with the body to which it belongs. See "Fieldwork, 2002–03: Tell el-Amarna, 2003," *Journal of Egyptian Archaeology* 89 (2003) 17–18, Pl. II.2.

ACKNOWLEDGEMENTS

I am most grateful to Dietrich Wildung, Director, Ägyptisches Museum, Berlin, for allowing me access to both the published and unpublished 'composite' elements excavated by Borchardt, and to discuss the latter in this paper. Also to Daphna Ben-Tor, Israel Museum, Jerusalem, who ferreted out the accession number of the Jerusalem forefoot on my behalf.

REFERENCES

Aldred, C., 1973, *Akhenaten and Nefertiti*. Viking Press, New York.

Arnold, D., 1996, *The Royal Women of Amarna. Images of Beauty from Ancient Egypt*. The Metropolitan Museum of Art, New York.

Borchardt, L., 1912, Ausgrabungen in Tell el-Amarna 1911/12. Vorläufiger Bericht, *Mitteilungen der Deutschen Orient-Gesellschaft zu Berlin*, 50, 1–40.

Borchardt, L.,1913. Ausgrabungen in Tell el-Amarna 1912/13. Vorläufiger Bericht, *Mitteilungen der Deutschen Orient-Gesellschaft zu Berlin*, 52, 1–55.

Hayes, W. C., 1959, *The Scepter of Egypt. A Background for the Study of the Egyptian Antiquities in The Metropolitan Museum of Art* II. *The Hyksos Period and the New Kingdom (1675–1080 B.C.)*. Rev. 1990, Metropolitan Museum of Art, New York.

Lacovara, P., Reeves, N. and Johnson, W. R., 1996. A Composite-statue Element in the Museum of Fine Arts, Boston, *Révue d'Égypologie*, 47, 173–176.

Lucas, A., 1962. *Ancient Egyptian Materials and Industries*. 4th edn., rev. J. R. Harris, Edward Arnold, London.

Michalowski, K, 1978, *Great Sculpture of Ancient Egypt*. Reynal & Co, New York.

Newman, R. and Serpico, M., 2000, Adhesives and binders, in P. T. Nicholson and I. Shaw (eds.), *Ancient Egyptian Materials and Technologies*, 475–501. University of Cambridge Press, Cambridge.

Nims, C. F., 1965, *Thebes of the Pharaohs. Pattern for Every City*. Elek Books, London.

Peet, T. E. and Woolley, C. L., 1923, *The City of Akhenaten* I. *Excavations of 1921 and 1922 at El-'Amarneh* (EES Memoir 38). Egypt Exploration Society, London.

Pendlebury, J. D. S., 1951, *The City of Akhenaten* III: *The Central City and the Official Quarters. The Excavations at Tell el-Amarna during the Seasons 1926–1927 and 1931–1936* (EES Memoir 44). Egypt Exploration Society, London.

Phillips, J. S., 1991, Sculpture Ateliers of Akhetaten: An examination of two studio-complexes in the City of the Sun-Disk, *Amarna Letters*, 1, 31–40.

Phillips, J.S., 1994, The Composite Sculpture of Akhetaten: Some Initial Thoughts and Questions, *Amarna Letters*, 3, 58–71.

Reeves, N. and Taylor, J. H., 1992, *Howard Carter before Tutankhamun*. Harry N. Abrams, New York.

Salah, M. and Sourouzian, H., 1987, *Official Catalogue the Egyptian Museum Cairo*. Philipp von Zabern, Mainz.

Samson, J., 1973, Amarna Crowns and Wigs. Unpublished pieces from statues and inlays in the Petrie Collection at University College, London, *Journal of Egyptian Archaeology*, 59, 47–59.

Samson, J., 1978, *Amarna, City of Akhenaten and Nefertiti*. Warminster, Aris & Phillips.

Wildung, D., 1995, Metamorphosen einer Königin: Neue Ergebnisse zur Ikonographie des Berliner Kopfes der Teje mit Hilfe der Computertomographie, *Antike Welt*, 26, 245–249.

Printed by Printforce, United Kingdom